HOW TO USE A HEAT PRESS

From Novice to Expert: Unleashing Your Creative Potential with Thermal Press Techniques

Fitzpatrick J. Thompkins

Copyright © 2024 by **Fitzpatrick J. Thompkins**

All rights reserved

No part of this publication may be reproduced, stored in a retrieval system, or transmitted, in any form or by any means, electronic, mechanical, photocopying, recording, or otherwise, without the prior written permission of the author.

The information in this ebook is true and complete to the best of our knowledge. All recommendation are made without guarantee on the part of author or publisher. The author and publisher disclaim any liability in connection with the use of this information.

Table of Contents

Introduction 5
   Overview of Heat Press Machines 6
   Benefits of Using a Heat Press 7

Chapter: 1 Types of Heat Press Machines 9
   Clamshell Heat Press 9
   Swing-Away Heat Press 11
   Draw Heat Press 13
   Sublimation Heat Press 15
   Comparison and Use Cases 17

Chapter: 2 Understanding Heat Press Components 19
   Platen (Heating Element) 19
   Pressure Adjustment 21
   Temperature Control 23
   Timer Control 25
   Silicone Pad and Teflon Sheets 27

Chapter: 3 Preparing for Heat Pressing 29
   Selecting the Right Materials and Fabrics 29
   Choosing Heat Press Accessories 31
   Design Preparation and Requirements 33
   Safety Measures and Workspace Setup 35

Chapter: 4 Basic Heat Press Techniques 37
   Setting Up Your Heat Press 37

Temperature, Pressure, and Time Settings	39
Placing Your Material and Transfer	41
Pressing and Removing the Transfer	43
Troubleshooting Common Issues	45
Chapter: 5 Advanced Heat Press Techniques	**47**
Layering Techniques for Multi-Colored Designs	47
Working with Different Types of Transfers	49
Tips for Sublimation Printing	51
Using Heat Press for Vinyl Applications	53
Chapter: 6 Maintenance and Troubleshooting	**55**
Daily and Periodic Maintenance Tips	55
Troubleshooting Common Problems	57
When to Seek Professional Help	59
Chapter: 7 Creative Projects and Ideas	**61**
Custom T-Shirts	61
Personalized Bags and Accessories	63
Home Decor Items	65
Unique Gift Ideas	67
Chapter: 8 Expanding Your Business with Heat Pressing	**69**
Marketing Your Products	69
Pricing Strategies	71
Scaling Production	73
Exploring New Markets	75
Chapter: 9 Health and Safety Considerations	**77**
Operating Safely	77
Ventilation and Air Quality	79

Handling Chemicals and Supplies   81
Conclusion   83

Introduction

In the heart of a bustling city stood a small, yet vibrant boutique called "Creative Threads." Its owner, Jamie, had long dreamed of expanding the shop's offerings beyond custom sewing to include personalized printed apparel. However, the journey from dream to reality was fraught with trial and error, particularly when it came to mastering the art of heat pressing.

The turning point came one autumn afternoon when Jamie discovered a book titled "How to Use a Heat Press" nestled among a stack of trade magazines. Skeptical but intrigued, Jamie began to leaf through the pages, and what was found within was nothing short of a revelation.

The book offered a comprehensive guide, starting from the very basics of understanding different heat press machines and their applications. It meticulously outlined the steps for preparing materials, choosing the right accessories, and setting up the workspace for optimal safety and efficiency. The detailed explanations on temperature, pressure, and time settings illuminated what had once seemed like guesswork. Moreover, the chapters dedicated to advanced techniques and creative projects sparked a wellspring of ideas in Jamie's mind, from custom T-shirts to personalized home decor items.

As days turned into weeks, Jamie applied the book's lessons, transforming "Creative Threads" into a hub of creativity and innovation. The boutique's new line of custom-printed apparel became a hit, attracting customers from far and wide. The secret to Jamie's success was no longer a mystery; it was the book that had been a constant companion through the learning curve.

But it wasn't just the technical knowledge that made the book invaluable. Its chapters on expanding a business, marketing products, and exploring new markets gave Jamie the confidence to dream bigger. Health and safety considerations ensured that the boutique remained a safe environment for creativity to flourish.

"Why should you buy this book?" Jamie would say to anyone curious about the secret behind the boutique's transformation. "Because it's more than just a manual; it's a roadmap to unlocking the full potential of your heat press and your creative business. It's filled with not only how-to's but also why's, ensuring you understand the principles behind each process. Whether you're a beginner or looking to refine your skills, this book guides you through each step with clarity and insight."

In the end, "How to Use a Heat Press" became more than a guide; it was a catalyst that transformed Jamie's boutique from a simple sewing shop into a beacon of custom apparel creativity. For Jamie and the countless others who followed in these footsteps, the book was an essential key to unlocking the door to endless possibilities.

Overview of Heat Press Machines

Heat press machines are pivotal in the world of custom apparel and merchandise, enabling users to transfer vivid designs onto various materials. These machines operate on a simple yet effective principle, using heat and pressure to imprint a design from a transfer paper onto your chosen medium, most commonly textiles. This technique has gained popularity not only for its efficiency but also for the durability and quality of the prints it produces.

There are several types of heat press machines, each designed to meet different needs and preferences. The clamshell design is renowned for its simplicity and space-saving benefits, making it a popular choice for beginners and those with limited workspace. Its upper heat platen opens vertically, resembling a clamshell. On the other hand, the swing-away design, where the upper platen swings away from the lower platen, offers more precision and safety, allowing a full view and access to the layout area without the risk of contact with the heat platen. The draw heat press features a design where the lower platen can be pulled out like a drawer, providing a safe and convenient way to layout your designs without having to work directly under the heated upper platen. Each type has its unique advantages, from ease of use to precision and safety, catering to different user requirements and work environments.

Beyond the basic operation of transferring designs, heat press machines offer a plethora of creative possibilities. They can work with a variety of materials and items, not just fabric, allowing for customization of a wide array of products such as mugs, hats, and plates, depending on the capabilities of the machine. The versatility of these machines opens up opportunities for businesses and hobbyists alike to explore a broad spectrum of custom merchandise.

Understanding the settings on a heat press machine is crucial for achieving optimal results. The temperature, pressure, and time must be carefully adjusted based on the material and the type of transfer being used. For instance, thicker materials and certain types of transfers may require higher temperatures or longer press times. The pressure adjustment is also vital, as too much pressure can cause the design to spread or blur, while insufficient pressure may lead to incomplete transfer.

Maintaining a heat press machine is relatively straightforward but essential for its longevity and performance. Regular cleaning of the platens and ensuring the machine is free from dust and debris can prevent many common issues. Additionally, the silicone pad and the Teflon sheets, which protect the material from direct heat and prevent sticking, should be kept in good condition.

In conclusion, heat press machines are a cornerstone in the creation of custom apparel and merchandise, offering efficiency, versatility, and quality in the transfer process. Whether for a small business, a large production operation, or personal projects, understanding the types, operation, and maintenance of these machines is essential for anyone looking to explore the world of heat transfer printing.

Benefits of Using a Heat Press

When exploring the realm of custom apparel and personalized items, the heat press emerges as an indispensable tool, offering a myriad of benefits to those who master its use. At the heart of its appeal is the ability to produce high-quality, durable prints on a variety of materials, ranging from t-shirts and hoodies to hats and bags. This versatility not only allows for a broad canvas on which creators can express their designs but also opens up numerous avenues for business opportunities in custom merchandise.

One of the most significant advantages of using a heat press is the precision and consistency it brings to the printing process. Unlike traditional printing techniques that may vary in quality and appearance, a heat press ensures that every item produced meets a high standard of quality, with vivid colors and clear, crisp designs that are resistant to fading and peeling over time. This reliability is crucial for businesses aiming to build a reputation for quality and for individuals who demand perfection in their personal projects.

The ease of use associated with heat presses is another compelling factor. With straightforward settings for temperature, pressure, and time, users can quickly learn the optimal conditions for different materials and transfer types. This ease of operation makes the heat press an accessible tool for beginners, while its advanced capabilities satisfy the demands of more experienced users. The learning curve is significantly flattened by resources such as comprehensive guides and tutorials, which detail the steps for creating professional-grade items.

Cost-effectiveness is yet another benefit of the heat press. For small businesses and hobbyists alike, the initial investment in a heat press can be quickly recouped through the production of custom merchandise. The ability to produce small batches on demand reduces waste and inventory costs, making it an economical choice for startups and established businesses aiming to expand their product offerings without significant risk.

The speed of production with a heat press cannot be understated. In a matter of seconds, designs can be transferred onto fabric or other materials, allowing for rapid turnaround times. This efficiency is invaluable in meeting customer demands, especially for businesses that offer personalized items. The swift production process also enables creators to experiment with designs and products without lengthy delays, fostering innovation and creativity.

Environmental considerations further underscore the benefits of using a heat press. Many heat transfer processes are cleaner and generate less waste than traditional screen printing and other methods. By using precise amounts of ink and reducing the need for water and chemicals, heat pressing is a more sustainable choice for those concerned with minimizing their environmental footprint.

Finally, the compact size and portability of many heat press models offer the convenience of setting up a printing station in limited spaces. This flexibility is essential for small studios, home businesses, or traveling vendors who require a portable solution for on-site customization.

In conclusion, the benefits of using a heat press, from the quality and durability of the prints to the cost-effectiveness and environmental advantages, make it an essential tool for anyone looking to explore the world of custom apparel and personalized items. Its ease of use, coupled with the support of detailed guides, ensures that users can quickly master the technique and unlock the full potential of their creative endeavors.

Chapter: 1 Types of Heat Press Machines

Clamshell Heat Press

The clamshell heat press stands as a cornerstone in the realm of heat transfer technology, distinguished by its unique design and functionality. This device, resembling a clamshell where the top platen opens upwards from the bottom, combines efficiency and simplicity, making it a favored choice for both beginners and seasoned professionals in the field of custom apparel and personalized items. Its design is not only space-saving but also user-friendly, offering a straightforward approach to applying heat transfers to a variety of materials.

The clamshell heat press is particularly lauded for its efficiency in setup and operation. The simplicity of its open-and-close mechanism facilitates a quick and easy process for placing and aligning garments and transfer materials. This design minimizes the physical effort required to operate the machine, making it accessible for users of all experience levels. Additionally, the compact footprint of the clamshell heat press makes it an ideal choice for those working in limited spaces, as it requires less room to operate compared to its swing-away counterparts.

When it comes to heat transfer applications, the clamshell design excels in delivering uniform pressure and consistent heat across the platen. This is crucial for achieving high-quality prints that are both vibrant and durable. The precision in heat application ensures that transfers are fully adhered to the fabric, minimizing issues such as peeling or fading over time. Furthermore, modern clamshell heat presses come equipped with digital controls that allow users to precisely set the temperature, pressure, and timing, tailoring the process to the specific requirements of the transfer material and the substrate.

For those looking to maximize the potential of their clamshell heat press, understanding the nuances of material compatibility and transfer types is key. For instance, certain heat-sensitive materials may require a careful balance of temperature and time to avoid damage, while thicker fabrics might need higher pressure for effective transfer. Experimenting with different materials and transfer techniques, such as sublimation, vinyl, or inkjet/laser transfers, can expand the range of products one can offer, from custom t-shirts and hoodies to hats, bags, and even hard goods like coasters and mouse pads.

Advanced users of clamshell heat presses can further enhance their craft by incorporating precision tools and accessories. Using pre-cut templates, laser alignment tools, or heat-resistant tape can improve the accuracy and consistency of placement, ensuring that designs are centered and straight. Additionally, employing various protective measures, such as Teflon sheets or silicone pads, can safeguard both the heat press and the materials from potential damage during the heat transfer process.

Maintenance and care of the clamshell heat press are also critical for sustaining its performance and longevity. Regularly cleaning the platen, checking the condition of the silicone pad, and ensuring the machine's hinges and opening mechanism are functioning smoothly can prevent common operational issues. By adhering to the manufacturer's guidelines for maintenance and promptly addressing any signs of wear or malfunction, users can ensure their clamshell heat press remains a reliable and effective tool for their heat transfer projects.

In conclusion, the clamshell heat press embodies a blend of simplicity, efficiency, and versatility in the realm of heat transfer printing. Its user-friendly design, combined with the capability to produce high-quality, durable prints, makes it an invaluable asset for anyone involved in custom apparel and personalized items. By leveraging advanced techniques, exploring a variety of materials and transfers, and maintaining the equipment with care, users can fully harness the potential of their clamshell heat press to expand their creative and business opportunities.

Swing-Away Heat Press

The swing-away heat press stands as a pivotal tool within the realm of custom printing, revered for its robustness, precision, and versatility. Distinguished by its design, where the upper heat platen swings away from the lower platen, this machine offers unparalleled access and safety, making it an ideal choice for intricate heat transfer projects. In delving into the advanced applications of a swing-away heat press, parallels can be drawn to cultivating the intricate and fascinating world of air plants, where precision, environment, and care foster growth and vibrancy.

At the heart of mastering the swing-away heat press is understanding its unique capability to evenly distribute heat and pressure, a critical factor when working with thick fabrics, multiple layers, or complex transfers. Similar to the meticulous care required for air plants to thrive—considering their need for bright, indirect light, and timely watering—the precision in heat press application ensures the high-quality adherence of designs without compromising the material's integrity.

To leverage the full potential of a swing-away heat press, it is essential to delve into the nuances of temperature control, pressure adjustments, and timing. These elements are akin to the environmental conditions essential for air plants, where slight variations can lead to significantly different growth outcomes. For instance, experimenting with slightly higher temperatures or longer press times can enhance the quality of certain transfers, much like adjusting the moisture levels or sunlight exposure can invigorate an air plant's growth.

Advanced users of swing-away heat presses often employ techniques such as pre-pressing the material to eliminate moisture and wrinkles, ensuring a flawless canvas for the transfer. This preparatory step mirrors the grooming of air plants, removing dead leaves and ensuring they are clean to promote healthy absorption of water and nutrients.

Another sophisticated strategy involves the use of silicone pads or Teflon sheets to protect the transfer and the material, particularly with delicate fabrics or high-detail designs. This approach is reminiscent of the careful placement of air plants within their environments, where strategic positioning away from direct sunlight can prevent damage while still fostering growth.

For those seeking to explore beyond traditional applications, the swing-away heat press offers the ability to work with a variety of materials and items that may not fit easily into clamshell models. This flexibility allows for the creation of a broad range of custom products, from intricate apparel designs to unique home décor items, much like the diverse array of vessels and settings in which air plants can be displayed.

Moreover, mastering the swing-away heat press for advanced applications requires an understanding of the specific requirements of different transfer types, such as vinyl, sublimation, or laser transfers. Each material and technique may require tailored settings and approaches, akin to the different care requirements of various air plant species, each with its unique aesthetic and environmental needs.

In conclusion, the swing-away heat press, with its sophisticated design and capabilities, offers a canvas for creativity and precision in the world of custom printing. By drawing parallels to the cultivation of air plants, we see the importance of understanding and adapting to the specific needs and nuances of both the medium and the method. Through careful attention to detail, experimentation, and a willingness to explore new techniques, users can unlock the full potential of their swing-away heat press, creating bespoke items that resonate with quality, durability, and artistic expression.

Draw Heat Press

The Draw Heat Press stands as a specialized tool within the realm of custom apparel and merchandise creation, embodying precision and versatility for enthusiasts and professionals alike. Its unique draw mechanism, where the lower platen slides out towards the operator, provides unparalleled ease of access and safety, reducing the risk of accidental burns during the placement and removal of items. This feature makes it particularly suitable for intricate heat transfer projects, including those involving air plants and other delicate designs that demand a gentle yet precise application of heat and pressure.

When leveraging a Draw Heat Press for advanced projects, such as creating custom merchandise adorned with air plant designs, there are several key considerations to ensure both the quality of the finished product and the longevity of the press itself. Understanding the specific attributes of the Draw Heat Press can significantly enhance the outcome of these projects, marrying the art of heat transfer with the delicate intricacies of botanical designs.

Firstly, the precision control over temperature, pressure, and timing is crucial for transferring detailed designs onto a variety of materials. The Draw Heat Press allows for fine-tuning of these parameters, ensuring that each transfer, whether it's a vivid air plant illustration or a complex botanical pattern, is executed flawlessly. For materials sensitive to heat or with a propensity for warping, the draw design minimizes the material's exposure to heat when setting up, a critical factor in maintaining the integrity of delicate designs.

Moreover, the slide-out feature of the Draw Heat Press affords operators the opportunity to closely inspect the layout on the platen before applying heat. This is particularly advantageous when working with complex designs or when precision placement is essential. It allows for a final adjustment without the risk of disturbing the item's positioning under a clamshell or swing-away model. This precise alignment capability is indispensable for achieving professional-grade results, especially when the designs feature the fine details characteristic of air plant imagery.

To further enhance the capabilities of the Draw Heat Press in advanced projects, operators can experiment with different types of heat transfer materials. From lightweight, translucent papers that mimic the ethereal quality of air plants to thicker, more textured materials that add depth and dimension to the design, the press's adjustable pressure settings make it adaptable to a wide range of transfer mediums. This adaptability is key for artists and creators who wish to push the boundaries of traditional heat press projects.

In addition to technical proficiency, maintaining the Draw Heat Press is paramount to ensure its longevity and performance.

Regular cleaning of the platen, careful inspection of the sliding mechanism for any signs of wear, and prompt replacement of silicone pads or Teflon sheets are all practices that preserve the press's functionality. Given the intricate designs and delicate applications often involved in advanced projects, the condition of the press can significantly impact the quality of the final product.

The Draw Heat Press, with its distinctive design and advanced capabilities, offers a sophisticated tool for creators looking to explore the intersection of botanical art and custom merchandise. Its precision, coupled with the operator's skill and creativity, opens up a world of possibilities for producing high-quality items that celebrate the delicate beauty of air plants and beyond. Whether for personal projects or commercial ventures, mastering the Draw Heat Press can elevate the standard of custom heat transfer items to new heights.

Sublimation Heat Press

In the realm of custom printing and design, the sublimation heat press stands out as a transformative tool, enabling creators to infuse materials with vivid, full-color images that last. This process involves transferring a design from sublimation paper onto a substrate, typically made of polyester or a polyester-coated material, using high heat and pressure. The ink transitions from a solid to a gas without becoming liquid, embedding itself into the material's fibers for a print that is incredibly durable and resistant to fading.

For those adept in the art of heat pressing, sublimation offers an exciting avenue for creativity and product diversity, but mastering this technique requires understanding its nuances. To achieve optimal results, it's crucial to select the right materials. Sublimation printing excels on polyester fabrics and polymer-coated substrates, where the ink can bond at a molecular level, ensuring the longevity and vibrancy of the design. Natural fibers like cotton do not accommodate sublimation ink as well, often leading to faded or washed-out images.

Temperature and time settings are paramount in sublimation printing. Each substrate may require a different heat setting, usually ranging between 380°F to 400°F, and a specific time frame to ensure the ink sublimates correctly. Maintaining these precise conditions is essential for a successful transfer, emphasizing the need for a heat press that allows for easy adjustment and monitoring of these parameters.

The pressure applied during the sublimation process also plays a critical role. Too little pressure can result in incomplete transfers, while too much pressure might cause the substrate to deform or the colors to bleed. Achieving a medium to high pressure, depending on the thickness and type of material, ensures a crisp, clear transfer of the design.

One of the most advanced tips for using a sublimation heat press effectively involves the concept of 'gassing out'. This phenomenon occurs when sublimated ink continues to activate and spread after the heat press is opened, which can blur or distort the design. To mitigate this, some professionals slightly open the heat press a few seconds before the timer ends, allowing some of the heat to dissipate and reduce the chance of gassing out.

Moisture management is another crucial consideration. Excess moisture in the paper or substrate can cause issues like color shifting, bleeding, or ghosting. Pre-pressing the substrate to eliminate any moisture and storing sublimation paper in a dry environment can help avoid these problems.

Experimentation with different materials and designs is a key aspect of mastering sublimation heat press techniques. The unique properties of different substrates, from hard surfaces like ceramics and metal to soft textiles, require adjustments and fine-tuning of the heat press settings. Testing and documenting these variations help in creating a comprehensive guide for future projects, ensuring consistency and quality in the production process.

In conclusion, mastering the sublimation heat press opens up a world of possibilities for custom printing, from apparel and accessories to home decor and beyond. By understanding and respecting the intricacies of the sublimation process—material selection, temperature and time settings, pressure adjustment, and moisture management—creators can elevate their projects to new heights, achieving professional-grade results that stand the test of time. This advanced knowledge, combined with a spirit of experimentation and refinement, is essential for anyone looking to grow their capabilities in the vibrant field of custom sublimation printing.

Comparison and Use Cases

Drawing a parallel between the nuanced world of growing air plants and mastering the art of using a heat press might seem unusual at first glance. However, both domains require a deep understanding of the environment, the right tools, and the nuances of care or technique to thrive and produce outstanding results. Just as air plant enthusiasts delve into advanced tips for growth, mastering the heat press involves exploring the diversity of machines and understanding their specific applications to enhance creativity and productivity.

The heat press landscape is populated with various models, each designed with particular use cases in mind, akin to the diverse species of air plants, each with its unique requirements for light, air, and moisture. Clamshell heat presses, for instance, are prized for their space-saving design, opening vertically much like the way Tillandsia might spread its leaves. This type makes it exceptionally suitable for straightforward projects where space is at a premium, similar to growing smaller air plants in compact spaces.

On the other hand, swing-away heat presses offer a higher level of precision and safety, allowing the upper platen to swing completely away from the lower platen. This design mimics the careful positioning of air plants in brighter light conditions to prevent scorching while ensuring even exposure. This type of press is ideal for projects requiring precise placement and heat application, such as high-detail transfers on t-shirts or fine fabrics, showcasing the parallel of needing precise environmental conditions for the flourishing of certain air plant varieties.

Draw heat presses introduce another dimension to the comparison, sliding out towards the user to allow for layout adjustments without the risk of contact with the heat source, akin to the cautious watering methods used for air plants to avoid water accumulation in their bases. This feature is particularly useful for projects that demand meticulous alignment and positioning, much like the careful cultivation techniques employed for more delicate air plant species.

Sublimation heat presses represent the pinnacle of versatility and are used to infuse dye directly into a wide range of materials, including ceramics and hard surfaces, not just fabrics. The process is reminiscent of the way some air plants absorb nutrients through their leaves from the air, requiring specific conditions to thrive. These presses are perfect for creating vibrant, full-color images on a variety of substrates, offering endless creative possibilities akin to the diverse visual displays air plants can offer through their life cycles.

In the realm of heat press use, the comparison extends into understanding the specific needs and best practices for different materials and projects. For example, using a clamshell press for a quick and efficient job on a bulk order of t-shirts mirrors the low-maintenance care of robust air plant species like Tillandsia xerographica. Conversely, employing a swing-away press for a complex, multi-layer vinyl project parallels the attentive cultivation of a delicate air plant species, requiring careful light and water conditions to thrive.

Moreover, just as air plant enthusiasts might experiment with different species and cultivation techniques to achieve a stunning display, heat press users can explore various machines, accessories, and methods to perfect their craft. From selecting the right pressure and temperature settings to using protective sheets and alignment tools, the nuances of mastering heat press use are vast and varied.

In conclusion, the journey to mastering the use of a heat press, much like growing air plants, is filled with learning, experimentation, and a deep appreciation for the nuances of each tool and project. By understanding the comparison and use cases of different heat press machines, users can select the perfect tool for their creative endeavors, ensuring a flourishing outcome that mirrors the beauty and diversity of a well-tended air plant display.

Chapter: 2 Understanding Heat Press Components

Platen (Heating Element)

The platen, or heating element, is a fundamental component of a heat press, playing a critical role in the transfer process of designs onto various materials. Its primary function is to provide consistent, controlled heat across its surface, which activates the adhesive on heat transfer materials, allowing images, text, or designs to be permanently affixed to the substrate. Understanding the nuances of the platen is crucial for anyone looking to master the art of using a heat press, as it directly impacts the quality and durability of the finished product.

Constructed typically from aluminum, the platen is engineered to ensure an even distribution of heat, which is vital for preventing cold spots that could lead to incomplete transfers. Aluminum is chosen for its excellent thermal conductivity, ensuring quick heating times and consistent temperature maintenance across the entire surface. This uniform heat distribution is crucial for achieving professional-grade results, regardless of the complexity or size of the design being transferred.

The surface of the platen is often coated with a non-stick material, such as Teflon, to prevent the transfer material from sticking to the platen and to ensure easy cleanup of any residual adhesive. This coating not only extends the life of the platen but also helps in preserving the integrity of the materials being pressed. The non-stick surface is especially beneficial when working with delicate fabrics or when attempting to remove misprinted materials without causing damage.

Temperature control is another critical aspect of the platen's functionality. Modern heat presses are equipped with digital temperature controls that allow users to precisely set and monitor the heat levels. This precision is necessary for working with a wide range of materials, each of which may require a specific temperature to achieve optimal adhesion without damaging the material. For example, heat-sensitive fabrics might require lower temperatures, whereas thicker, more durable materials might need higher settings to ensure a successful transfer.

The size and shape of the platen are also important considerations, as they determine the size of the designs that can be transferred and the types of items that can be accommodated by the heat press. Standard flat platens are suitable for a broad range of applications, including t-shirts, tote bags, and flat fabric surfaces. However, specialized platens are available for pressing onto hats, mugs, plates, and other non-flat items, enabling users to expand their range of custom merchandise.

Maintenance of the platen is straightforward yet essential for maintaining the quality of transfers and extending the lifespan of the heat press. Keeping the platen clean and free of adhesive residue ensures that each transfer is clean and sharp. Regular checks for any signs of wear or damage, particularly to the non-stick coating, can prevent unexpected issues during the heat press operation.

In summary, the platen or heating element is a cornerstone of the heat press's functionality. Its design and maintenance are critical for ensuring even heat distribution, precise temperature control, and versatility in the types of materials and items that can be customized. As such, a deep understanding of the platen's characteristics and proper care is indispensable for anyone keen on leveraging the full potential of a heat press for creating high-quality, durable custom merchandise.

Pressure Adjustment

Understanding pressure adjustment is fundamental to mastering the art of using a heat press. As a crucial component of the heat pressing process, pressure adjustment plays a significant role in ensuring the quality and durability of the transferred designs onto various materials. Whether working with delicate fabrics or more resilient items, the ability to precisely control the pressure applied by the heat press can make the difference between a flawless finish and a failed project.

Pressure adjustment in heat press machines allows the operator to modify the force with which the press's plates come together during the transfer process. This adjustment is critical because different materials and transfer types require varying levels of pressure to achieve optimal adhesion and quality. For instance, thicker materials might necessitate higher pressure to ensure the design adheres well, while thinner, more delicate fabrics require a lighter touch to prevent damage.

The mechanics of pressure adjustment vary among heat press models, but the principle remains consistent: to provide an even, consistent application of pressure across the material. Clamshell, swing-away, and draw heat press designs each incorporate mechanisms for adjusting pressure, typically through knobs or digital controls that allow the user to increase or decrease the pressure according to the specific needs of the project at hand.

Achieving the right pressure setting is a balance between the type of material, the transfer medium (such as vinyl, inkjet paper, or sublimation paper), and the specific requirements of the design to be transferred. For example, a thicker design may require more pressure to ensure all parts of the design adhere evenly to the fabric, while a very thin or delicate design might need less pressure to avoid blurring or bleeding edges.

The importance of uniform pressure distribution cannot be overstressed. Uneven pressure can lead to parts of the design not transferring correctly, resulting in areas that are faded, missing, or inconsistently adhered. This is why high-quality heat presses are designed to ensure that pressure is evenly distributed across the platen, regardless of the pressure setting. Operators must be familiar with their machine's particular pressure adjustment controls and how to gauge the pressure applied to ensure consistent, high-quality results.

Moreover, mastering pressure adjustment involves understanding the feedback from the heat press and the materials being used. This might include recognizing signs of too much pressure, such as the substrate material stretching or the transfer medium bleeding, or too little pressure, evident by incomplete transfer or peeling of the design after washing. Experience, coupled with careful observation and adjustment, helps operators develop a feel for the correct pressure settings for a wide range of projects.

In the context of learning how to use a heat press, understanding and mastering pressure adjustment is essential. It's not merely about following a set of instructions but developing a nuanced understanding of how pressure affects different materials and designs. With practice and attention to detail, operators can use pressure adjustment to their advantage, producing professional-quality prints that stand the test of time.

In conclusion, pressure adjustment is a critical aspect of the heat pressing process, affecting the quality, durability, and overall success of the printed design. By understanding how to accurately adjust and apply pressure, users can ensure that each project from their heat press comes out looking its best.

Temperature Control

Temperature control is a cornerstone feature in the operation of heat press machines, playing a pivotal role in the successful transfer of designs onto various substrates. Understanding and mastering temperature settings are essential for anyone looking to achieve high-quality, durable prints through heat pressing. This component's significance is magnified when considering the diverse range of materials and transfer types used in the process, each requiring specific heat settings to ensure optimal adhesion and color fidelity.

At its core, temperature control in a heat press involves adjusting the machine to the precise heat level recommended for the specific transfer paper, vinyl, or other materials being used. The heat press's ability to accurately reach and maintain the set temperature is crucial for ensuring that the heat-sensitive adhesives in transfer papers and vinyls are activated properly, allowing them to bond effectively with the substrate. This bond is what gives the printed design its durability, making it resistant to washing and wear.

Different materials and transfer methods have unique temperature requirements. For example, lightweight fabrics like polyester might require lower temperatures to prevent scorching, while thicker materials like cotton can withstand higher heats. Sublimation transfers, used to print on polyester and polymer-coated substrates, typically require high temperatures to turn the solid dye particles directly into gas, embedding the dye into the fabric. In contrast, transfers designed for light or dark cotton may need different temperature settings to achieve the best results.

The heat press's temperature control mechanism often includes a digital or analog thermostat that allows the user to set the temperature precisely. Digital models offer the advantage of more accurate temperature settings and easier monitoring, often displaying both the set temperature and the current temperature of the platen. This feature is invaluable for avoiding the heat variances that can lead to incomplete transfers or damaged materials.

Moreover, understanding the interplay between temperature and time is critical for effective heat pressing. Higher temperatures may reduce the time needed for a transfer, but they also increase the risk of damaging sensitive materials. Conversely, lower temperatures may require a longer pressing time, which can also affect the outcome of the print. This delicate balance is why detailed heat press guides and manufacturers' recommendations are indispensable tools for operators, providing the starting points for experimentation and fine-tuning.

The quality of a heat press's temperature control system directly impacts the machine's overall performance and the longevity of its prints. Fluctuations in temperature during the pressing process can result in uneven transfers, where parts of the design may not adhere properly or appear faded. Consistent temperature maintenance ensures uniformity across the entire design, leading to professional-grade results even in a home setting.

Regular maintenance of the heat press, including the cleaning of the heating element and the inspection of the thermostat, ensures the accuracy of temperature settings over time. This preventative care is essential for prolonging the life of the machine and maintaining the quality of its output.

In conclusion, temperature control is more than just a setting on a heat press; it is a fundamental principle that influences every aspect of the heat transfer process. From the selection of materials and transfer papers to the fine-tuning of pressing times, the ability to precisely control temperature allows for the customization and creation of high-quality, durable designs. Mastery of this component is a crucial step for anyone looking to excel in the use of a heat press, whether for personal projects or commercial production.

Timer Control

In the intricate dance of applying heat and pressure to transfer designs onto various materials, the timer control of a heat press stands out as an essential component, guiding users through each step of the process with precision and reliability. Understanding how to effectively utilize the timer is crucial for anyone seeking to master the art of using a heat press, as it directly influences the quality and durability of the final product.

The timer control on a heat press serves as the conductor of the operation, dictating the duration for which heat and pressure are applied to the material and transfer. This is not merely a matter of convenience but a critical factor in ensuring that designs are flawlessly transferred without being under or over-exposed to heat, which can lead to fading, peeling, or even burning of the material.

Setting the timer appropriately requires an understanding of the specific requirements of the material being used, the type of transfer, and the desired outcome. For instance, thicker fabrics may require a longer press time compared to thinner ones, and certain types of transfers might need more or less time to properly adhere to the material. The timer control allows for adjustments in small increments, enabling users to fine-tune the process to achieve optimal results.

One of the key advantages of utilizing the timer control effectively is the consistency it brings to the production process. By adhering to the recommended time settings for each material and transfer type, users can produce items that are consistently high in quality, enhancing both the reputation of the business and the satisfaction of the customers. This consistency is especially important when producing multiple items in a batch, ensuring that each piece meets the same high standards.

Moreover, the timer control plays a pivotal role in the efficiency of the heat pressing process. By precisely managing the time each item spends under the press, users can streamline their workflow, reducing the likelihood of errors and rework. This efficiency is critical in a commercial setting, where time is of the essence, and productivity directly impacts profitability.

Advanced heat press models often feature digital timer controls with user-friendly interfaces, making it easier than ever to set precise timings. These digital controls may also offer additional functionalities, such as audible alarms or alerts, to signal when the pressing cycle is complete. This feature is particularly useful in busy environments, where the operator may be multitasking and needs a reminder to release the press and remove the item.

In addition to enhancing the quality and consistency of the finished products, proper use of the timer control contributes to the longevity of the heat press itself. By preventing overexposure to heat, it helps maintain the integrity of the heating element and other critical components, reducing wear and tear and extending the life of the machine.

In summary, the timer control is a fundamental component of a heat press that significantly influences the success of the heat transferring process. Mastery of this feature enables users to achieve precise, consistent, and high-quality results, optimizing both the operation of the heat press and the durability of the printed items. Whether for personal projects or commercial production, an in-depth understanding of the timer control is a cornerstone of proficient heat press use, paving the way for creative expression and business growth in the realm of custom apparel and personalized products.

Silicone Pad and Teflon Sheets

In the intricate process of mastering heat press operations for custom apparel and merchandise, two components play pivotal roles in ensuring the quality and longevity of printed items: silicone pads and Teflon sheets. These components, though simple in appearance, are fundamental in navigating the heat press landscape, offering a blend of protection, heat distribution, and non-stick surfaces that contribute to the overall success of heat pressing projects.

Silicone pads, nestled beneath the material being printed, serve as a resilient, heat-resistant barrier. They play a crucial role in evenly distributing the heat and pressure applied during the pressing process. This even distribution is essential for consistent transfers, ensuring that every part of the design adheres properly to the fabric without areas of under or over-application. Silicone pads also provide a slight cushioning effect, accommodating slight variances in material thickness and surface imperfections, which can be particularly beneficial when working with garments of varying textures and weaves. Over time, these pads may wear down or become compressed, particularly in high-use areas, necessitating periodic inspection and replacement to maintain optimal performance.

Teflon sheets, on the other hand, are placed on top of the transfer paper and material during the pressing process. The non-stick property of Teflon (PTFE) is its most celebrated feature, preventing the transfer from sticking to the heat plate, which could ruin both the design and the garment. Moreover, Teflon sheets help in protecting the material from direct contact with the heated plate, reducing the risk of scorching or heat damage that can occur with sensitive fabrics. This protective layer ensures that the heat is applied smoothly and consistently across the transfer, aiding in achieving a polished and professional finish.

The utility of Teflon extends beyond its non-stick properties; it also acts as a barrier against ink bleed, which can happen when high temperatures cause the dye to spread outside the intended design area. By controlling this spread, Teflon sheets help in maintaining the sharpness and clarity of the design, ensuring that the final product reflects the creator's vision with precision.

Using silicone pads and Teflon sheets in conjunction introduces an additional layer of versatility to heat pressing techniques. For instance, adjusting the silicone pad thickness or layering Teflon sheets can modify the pressure and heat experienced by the material, allowing for fine-tuning the process to suit a wide range of materials and transfer types. This adaptability is invaluable in a field where innovation and customization are key.

Understanding the functions and benefits of silicone pads and Teflon sheets illuminates their indispensable roles in the heat pressing process. These components not only safeguard the materials and equipment but also enhance the quality and durability of the printed items. As such, maintaining these components in good condition and employing them correctly are fundamental practices for anyone looking to achieve professional-grade results in their heat press projects. Whether for a hobbyist exploring personal projects or a business producing commercial merchandise, the knowledge and application of these heat press components are essential chapters in the broader narrative of mastering heat press technology.

Chapter: 3 Preparing for Heat Pressing

Selecting the Right Materials and Fabrics

Selecting the right materials and fabrics is a critical step in the heat pressing process, pivotal for achieving high-quality, long-lasting results. This selection process is influenced by the compatibility of materials with the heat press technique, the desired outcome of the project, and the specific characteristics of the materials and fabrics in question.

At the core of material selection is understanding the heat sensitivity of different fabrics. Natural fibers like cotton and linen are renowned for their heat tolerance, making them ideal candidates for heat pressing. These materials can withstand high temperatures, allowing for the transfer of designs without damaging the fabric. On the other hand, synthetic fibers such as polyester, nylon, and spandex require more careful consideration due to their heat sensitivity. While these materials can also be used for heat pressing, it's essential to adjust the temperature settings and pressing time to prevent melting or warping.

Blended fabrics, which combine natural and synthetic fibers, present their own set of challenges and advantages. They often offer the best of both worlds—durability and ease of printing seen in natural fibers, with the added strength and elasticity of synthetics. However, the key to successfully heat pressing blended fabrics lies in finding the right balance of heat and pressure that suits both types of fibers in the blend.

Beyond the type of fabric, the color and weight of the material also play significant roles in the heat pressing process. Light-colored fabrics typically showcase vibrant, true-to-color transfers without the need for adjustments. Dark or brightly colored fabrics, however, may require the use of special transfer papers or additional layers to ensure that the design remains visible and vibrant after pressing. The weight of the fabric influences the pressure setting on the heat press; heavier fabrics may require more pressure to ensure the design adheres properly, whereas delicate fabrics need a gentler approach.

The intended use of the finished product is another crucial factor in selecting materials. For items that will be washed frequently, such as clothing and linens, choosing a fabric that can maintain the integrity of the transfer through numerous wash cycles is vital. Durability becomes a key consideration, emphasizing the importance of selecting high-quality fabrics and transfer materials that can withstand wear and tear.

Finally, the finish and texture of the fabric can affect the appearance and feel of the final product. Smooth fabrics provide a clean, even surface that can showcase detailed designs with clarity. Textured materials, while offering a unique tactile experience, may not be suitable for intricate designs due to the potential for uneven transfer.

In sum, selecting the right materials and fabrics for heat pressing is a nuanced process that requires a thorough understanding of the properties of different fabrics, the specifics of the heat press technique, and the objectives of the project. By carefully considering factors such as fabric type, color, weight, intended use, and texture, creators can ensure that their projects not only look professional but also stand the test of time. This careful selection process underpins the success of heat pressing projects, turning creative visions into tangible, high-quality products.

Choosing Heat Press Accessories

Choosing the right accessories for your heat press is a pivotal step in preparing for successful heat pressing projects. The accessories not only enhance the functionality of your heat press but also ensure the quality and durability of your prints, while protecting the materials you work with. Understanding the variety and purpose of these accessories can significantly impact the final outcome of your products.

At the forefront of essential heat press accessories are Teflon sheets and silicone pads. Teflon sheets are used to protect the upper heat platen from ink bleed and to ensure an even heat distribution across the transfer. This is crucial for achieving a smooth, professional finish on your prints. Silicone pads, on the other hand, are placed on the lower platen to provide a firm but resilient surface for pressing. They help in accommodating items of various thicknesses and in absorbing excess pressure, thereby preventing marks or indentations on the material being printed.

Heat-resistant tape is another indispensable accessory, particularly when working with materials that are prone to shifting or sliding during the pressing process. This tape secures the transfer paper in place, ensuring precise alignment and preventing ghosting or blurring of the design. Being heat-resistant, it does not leave any residue on the fabric or transfer, making for a clean, crisp print.

For those looking to expand their range of printable items, specialty platens are a worthy investment. These are interchangeable platforms that can be attached to the heat press to accommodate items of different shapes and sizes, such as hats, sleeves, shoes, or bags. Specialty platens allow for more precise positioning and pressure application, opening up a world of possibilities for custom merchandise.

Another key accessory to consider is the foam pillow. Foam pillows are used to raise the printing area of a garment or item, especially when working around seams, zippers, or buttons. By ensuring an even surface, foam pillows help in achieving consistent prints across challenging items, making them invaluable for a diverse product offering.

For businesses or individuals aiming to specialize in sublimation printing, a sublimation paper specifically designed for this process is essential. Sublimation paper facilitates the transfer of designs onto materials with a polyester coating or blend, enabling vibrant, full-color images that are embedded into the fabric for a lasting finish.

Protective gloves and garments should not be overlooked when choosing heat press accessories. The high temperatures involved in heat pressing can pose a risk of burns. Heat-resistant gloves provide a safeguard against these risks, allowing for safe handling of materials and transfers. Similarly, wearing protective garments can shield against accidental contact with the heated platens.

Lastly, investing in a good quality heat press cleaning kit is vital for the maintenance and longevity of your machine. Regular cleaning ensures the platens remain free of residual inks, adhesives, and other contaminants that can affect print quality. A cleaning kit typically includes non-abrasive cloths, mild cleaning solutions, and brushes designed to safely clean the machine without damaging its components.

In preparing for heat pressing, the thoughtful selection of accessories is as important as choosing the right heat press itself. Each accessory serves a specific purpose, from enhancing the quality of the prints to expanding the range of products you can offer. By carefully considering the needs of your projects and the capabilities of your heat press, you can assemble a set of accessories that will elevate your printing process, ensure safety, and support your creative vision.

Design Preparation and Requirements

In the intricate process of heat pressing, the preparation of designs and understanding their requirements are pivotal steps that determine the quality and success of the final product. This stage involves a careful consideration of various factors, from the choice of artwork to the selection of materials, all tailored to ensure that the finished items meet both aesthetic and practical standards.

The foundation of successful design preparation begins with the selection of the artwork. High-resolution images and vector graphics are preferred due to their scalability and clarity. These formats allow for adjustments in size without losing quality, ensuring that the design remains crisp and clear when transferred onto a fabric or other substrates. It's crucial for the artwork to be mirrored or reversed before printing, especially when working with heat transfer papers, as this ensures the design appears correctly oriented once applied.

Color management is another critical aspect of design preparation. The colors on a computer screen may not always translate accurately onto the printed material due to differences in color spaces between digital devices and physical prints. Utilizing color correction tools and consulting color charts can help in achieving colors that are as close as possible to the intended design. This step is particularly important for designs that require precise color matching, such as logos or branded materials.

The choice of material also plays a significant role in design preparation. Different fabrics and substrates react differently to heat and pressure, and selecting the right transfer paper or vinyl is crucial. For example, certain materials are better suited for light or dark fabrics, while others are designed specifically for hard surfaces. Understanding the compatibility between the material and the transfer medium ensures not only the durability of the print but also its visual appeal.

Preparing the artwork also involves considering the size and placement of the design. The dimensions of the design should be tailored to fit the item it will be transferred onto, whether it's a t-shirt, bag, or any other merchandise. Placement is key to achieving the desired look, whether it's centered, off-center, or aligned to specific margins. A mock-up on a digital template or using a physical sample can provide a visual reference to ensure the design aligns with the intended aesthetic.

In addition to these considerations, the technical requirements of the heat press machine must not be overlooked. Each machine has its own specifications for temperature, pressure, and time settings, which can vary depending on the material and the type of transfer being used. Familiarizing oneself with these settings and adjusting them according to the manufacturer's recommendations or through trial and error is essential for achieving optimal results.

Lastly, testing the design on a sample material before proceeding with the final application is a prudent practice. This not only allows for assessing the quality of the transfer but also provides an opportunity to make necessary adjustments to the design or the heat press settings. It's a critical step for avoiding costly mistakes and ensuring the finished product meets the expected standards of quality.

In conclusion, the preparation of designs and understanding their requirements are foundational to mastering the art of heat pressing. This process demands attention to detail, from the selection and preparation of the artwork to the choice of materials and the adjustment of heat press settings. By meticulously addressing each of these aspects, individuals can enhance the quality of their creations, turning simple items into personalized masterpieces that stand the test of time.

Safety Measures and Workspace Setup

Preparing for heat pressing involves not just understanding how to operate the machine but also setting up a safe and efficient workspace. The nature of heat press machines, operating at high temperatures and requiring manual operation, necessitates a keen focus on safety measures to prevent accidents and ensure a smooth workflow.

Creating a conducive environment for heat pressing begins with selecting an appropriate space. This area needs to be well-ventilated to dissipate heat and fumes that may emanate from the materials being pressed, especially when working with synthetic fabrics or using certain types of transfers that might release odors or chemicals when heated. Adequate ventilation not only contributes to a safer working environment but also enhances the comfort of the operator.

The surface on which the heat press is placed must be stable, level, and capable of withstanding the weight of the machine. Heat presses, particularly the larger models, can be quite heavy and applying pressure during the pressing process generates additional force. A sturdy table or workbench is essential to prevent the machine from moving or tipping over, which could cause injuries or damage.

Spacing around the heat press is another critical consideration. Operators need sufficient room to move freely and safely, particularly when working with larger items such as banners or blankets. This space is also necessary for laying out materials, prepping items for pressing, and cooling them down post-pressing. Clutter should be minimized to reduce the risk of accidents, such as tripping or accidentally placing flammable materials too close to the heat source.

Personal protective equipment (PPE) plays a crucial role in operator safety. Heat-resistant gloves are a must to protect hands from burns when handling hot materials or adjusting the press. Safety glasses or goggles provide eye protection against possible splatter from inks or other materials. Depending on the specific materials and inks being used, a mask or respirator might be necessary to avoid inhalation of potentially harmful fumes.

In terms of electrical safety, ensuring the heat press is properly grounded is paramount to prevent electrical shocks. The electrical outlet used should match the machine's power requirements, and the use of extension cords should be avoided if possible, as they can be a fire hazard if not rated for the machine's power draw.

Regular maintenance of the heat press is also a safety measure. Keeping the machine clean, checking for worn or damaged parts, and following the manufacturer's guidelines for maintenance can prevent malfunctions that might lead to accidents. This includes checking the wiring and the condition of the heating element, platen, and silicone pad for signs of wear or damage.

Finally, emergency preparedness should not be overlooked. Having a first aid kit readily available, knowing the location of the nearest fire extinguisher, and familiarizing oneself with the quickest route out of the workspace in case of emergency are all vital safety protocols.

In summary, safety measures and workspace setup are foundational aspects of preparing for heat pressing. A well-ventilated, spacious, and organized workspace, combined with the use of personal protective equipment and adherence to electrical safety standards, sets the stage for a safe and productive heat pressing operation. Regular maintenance and emergency preparedness further ensure that the use of a heat press is both effective and safe, allowing for creativity to flourish without compromising the well-being of the operator.

Chapter: 4 Basic Heat Press Techniques

Setting Up Your Heat Press

Setting up your heat press correctly is the cornerstone of mastering basic heat press techniques, an essential skill for anyone looking to create high-quality, custom apparel or items. This process involves several critical steps, from selecting the right space for your operations to adjusting the machine's settings for optimal results. Each step, when executed with care and precision, ensures that the end product meets the desired standards of quality and durability.

Firstly, choosing an appropriate workspace is paramount. The area should be well-ventilated to prevent the buildup of heat and should have a stable, flat surface to accommodate the heat press machine. Adequate space not only ensures safety but also allows for efficient movement and organization of materials and designs. It's important to have all necessary supplies within reach, including heat transfer paper, fabrics, and any protective coverings like silicone pads or Teflon sheets.

Once the workspace is established, the next step is to familiarize yourself with the heat press machine. This involves understanding its components, such as the platen, where heat and pressure are applied; the pressure adjustment knob, which controls the force applied to the material; and the temperature and time settings, which vary depending on the material and transfer type. Many heat press machines also feature digital displays and controls, making it easier to set precise temperatures and timers.

Before beginning any project, it's crucial to test the heat press. This involves heating the machine to the recommended temperature for the material and transfer paper being used. A test press on a scrap piece of fabric or material similar to your project can help you adjust the pressure and temperature settings if needed. This step helps avoid common issues like overheating, which can cause transfers to burn or not adhere properly.

The correct placement of your material and design is next. The fabric or item should be laid flat on the platen, ensuring there are no wrinkles or folds. The transfer paper, with your design, should then be placed in the desired location on the material, often with a protective layer on top to prevent scorching. Ensuring proper alignment at this stage is key to achieving professional-looking results.

The heat press operation is relatively straightforward but requires attention to detail. Once the machine is set to the correct temperature and pressure, and the timer is adjusted for the specific project, the press is closed over the material and design.

After the timer goes off, indicating the completion of the press, the machine is opened, and the material is carefully removed. Depending on the type of transfer, a cooling period may be necessary before peeling off the transfer paper to reveal the final design.

In addition to these steps, maintaining a clean and organized workspace and heat press machine is crucial for consistent results. Regular cleaning of the platen and other components of the heat press ensures that residue from previous projects does not transfer to new items. Keeping a log of successful temperature, pressure, and timer settings for different materials and designs can also save time and reduce errors in future projects.

Setting up your heat press correctly, from preparing the workspace to conducting test presses and adjusting settings, lays the foundation for mastering basic heat press techniques. This process not only ensures the quality and durability of your custom creations but also enhances efficiency and productivity, whether you're producing items for personal enjoyment or commercial purposes.

Temperature, Pressure, and Time Settings

Mastering the use of a heat press involves understanding the critical relationship between temperature, pressure, and time settings. These three elements are foundational to achieving high-quality, durable prints on various materials. Properly balancing these settings ensures that the heat transfer process results in vibrant, crisp designs that last through countless washes and wear.

Temperature is the cornerstone of the heat press operation. Different materials and transfer types require specific heat levels to effectively bond the design to the substrate. For instance, a cotton t-shirt might require a different temperature setting compared to a polyester blend fabric. Heat-sensitive materials demand lower temperatures, while thicker, more resilient materials can withstand higher heat. The key is to follow the heat transfer material manufacturer's recommendations closely to avoid damaging the substrate or the transfer. Temperatures typically range from 300°F to 400°F, depending on the material and transfer type.

Pressure is equally vital in the heat pressing process. It ensures the even transfer of the design onto the material's surface. The amount of pressure needed can vary significantly based on the type of material and the specific transfer. For example, thicker materials may require more pressure to ensure the design adheres well to every fiber, while delicate materials need lighter pressure to prevent damage. Heat press machines often categorize pressure settings as light, medium, or heavy, and fine-tuning this balance is crucial for optimal results. Uniform pressure across the entire platen is essential to prevent parts of the design from not transferring correctly.

Time settings determine how long the material and transfer are exposed to the set temperature and pressure. This aspect is critical because too little time might result in an incomplete transfer, while too much time can burn the material or degrade the quality of the transfer. Time settings generally range from a few seconds for light transfers to up to a minute for more complex or layered designs. The exact time required can depend on the temperature and pressure settings, as well as the specific characteristics of the material and transfer type being used.

The interplay between temperature, pressure, and time is a delicate balance that requires attention and experimentation. Manufacturers of heat press machines and transfer materials often provide recommended settings, but these should be viewed as starting points. Variations in equipment, environment, and specific materials might necessitate adjustments.

Experimentation plays a crucial role in mastering these settings. Keeping a log of successful settings for different material and transfer combinations can be incredibly helpful. This record-keeping practice saves time in future projects and helps in troubleshooting when issues arise. Additionally, conducting test presses on scrap material before committing to the final product is always advisable to ensure the settings are perfectly tuned for the task at hand.

Understanding and mastering the temperature, pressure, and time settings are fundamental to unlocking the full potential of a heat press. This knowledge not only ensures the production of high-quality, durable custom apparel and items but also significantly enhances the efficiency and satisfaction of the heat pressing experience. Whether for personal projects or commercial production, the ability to skillfully manipulate these settings is what differentiates the novice from the expert in the world of heat press printing.

Placing Your Material and Transfer

Placing your material and transfer correctly in the heat press is a fundamental step in the process of creating high-quality custom apparel and items. This phase is crucial because it directly impacts the final outcome, determining the precision, adherence, and overall appearance of the design on the product. Mastering this technique ensures that each item produced meets the desired standards of quality and durability.

The process begins with the preparation of the material and the transfer. The material, whether it's a t-shirt, bag, or any other fabric-based item, should be pre-treated if necessary and laid out flat to ensure there are no wrinkles or moisture. Moisture can affect the heat transfer process, leading to uneven application or adherence issues. Similarly, wrinkles in the fabric can cause the design to be applied improperly, resulting in a flawed final product. Pre-pressing the material for a few seconds in the heat press can remove moisture and wrinkles, providing a smooth surface for the transfer.

Selecting the appropriate transfer paper is just as important as preparing the material. The type of transfer paper depends on the material of the product and the type of heat press being used. For instance, different papers are used for light and dark fabrics, and some are specifically designed for sublimation printing. Ensuring the correct match between the transfer paper and the material guarantees that the ink adheres properly and that the colors appear as intended.

Once the material and transfer paper are ready, the next step is to position the transfer paper correctly on the material. This positioning is vital for ensuring that the design is centered, straight, and placed at the correct spot on the material. Many heat press machines come with laser alignment tools or other aids to help in accurately placing the transfer. If such tools are not available, a simple ruler or tape measure can be used to align the transfer manually. It's important to take the time to ensure the alignment is perfect to avoid off-center designs or other mistakes that could render the product unsellable.

After positioning the transfer paper, it's time to move on to the heat press. The material, now with the transfer paper placed on top, should be carefully laid onto the heat press platen. The side of the transfer paper with the design or ink should be facing down against the material, ready to be transferred. At this point, it's crucial to ensure that the material is flat on the platen, without any folds or overlaps that could affect the transfer.

The next step involves setting the heat press to the correct temperature, pressure, and time settings for the material and transfer type. These settings vary depending on the material of the product, the type of ink used in the transfer, and the specific requirements of the transfer paper. Following the manufacturer's recommendations for these settings is essential for achieving the best results.

Once the heat press is set up, the pressing process can begin. Lowering the heat press onto the material and transfer paper applies heat and pressure, activating the transfer process. The heat causes the ink on the transfer paper to liquefy and permeate the material, while the pressure ensures that the transfer is evenly and firmly applied across the entire design.

After the specified time has elapsed, the heat press can be lifted, and the material removed. If using hot-peel transfer paper, the backing should be peeled off immediately while it's still hot. For cold-peel transfer papers, the material should be set aside to cool before the backing is removed. Once the backing is peeled away, the transfer process is complete, and the design is permanently affixed to the material.

In conclusion, placing your material and transfer correctly is a critical step in using a heat press effectively. Attention to detail during this phase ensures that the final product is of high quality, with a clear, vibrant design that meets or exceeds expectations. Mastery of this technique is essential for anyone looking to produce professional-grade custom apparel and items.

Pressing and Removing the Transfer

The process of pressing and removing the transfer is a pivotal stage in the heat press operation, where the precision of your technique can significantly influence the quality and longevity of the final product. This step involves the careful application of heat and pressure to transfer your design onto the chosen material, followed by the critical moment of peeling away the transfer paper or film to reveal the finished design. Mastering this phase is essential for anyone looking to achieve professional-grade results in their custom apparel or personalized items.

The initial aspect to consider in the pressing phase is the preparation of your material and the positioning of your transfer. Ensuring that the material is smooth, free of wrinkles, and correctly aligned on the heat press platen sets the foundation for a successful transfer. The placement of your transfer paper or film, with the design facing down onto the material, must be precise to avoid misalignment or partial designs.

Once the material and transfer are correctly positioned, the heat press is closed to initiate the pressing process. The temperature, pressure, and time settings should be meticulously adjusted according to the type of material and transfer paper being used. These parameters are crucial; too high a temperature or excessive pressure can damage the material or the design, while insufficient heat or pressure can lead to a transfer that is poorly adhered or incomplete. The recommended settings are often provided by the transfer paper manufacturer and should be followed closely to ensure optimal results.

After the designated time has elapsed, the press is opened, and this is where the process diverges based on the type of transfer paper used. With some transfers, immediate peeling (hot peel) is required, while others necessitate waiting until the material cools down (cold peel). Recognizing the correct method to use is essential, as premature or delayed peeling can affect the quality of the design transfer. A smooth, steady motion is recommended when removing the transfer paper or film, ensuring that the entire design has adhered to the material before fully removing the backing.

Attention to detail during the peeling phase is paramount. Observing the design as it separates from the transfer paper allows for quick identification of any areas that may not have fully adhered. In such cases, reapplying the press for a brief period can sometimes salvage the transfer. However, caution is advised to avoid overheating or applying excessive pressure that could compromise the integrity of the design.

After the transfer paper is removed, some materials and transfers may require a post-pressing phase without the transfer paper, often with a protective silicone sheet or parchment paper. This additional step can help to set the design, ensuring it is fully bonded to the material and offering an extra layer of protection to enhance durability.

The successful pressing and removal of the transfer are integral to creating high-quality, durable custom apparel and personalized items. This process, while seemingly straightforward, requires a keen understanding of the interplay between material, heat, pressure, and timing. Mastery of this phase, as outlined in comprehensive guides on using a heat press, empowers users to produce professional-level results consistently, paving the way for exploration and innovation in custom design.

Troubleshooting Common Issues

Troubleshooting common issues is an essential skill when working with heat press machines, ensuring smooth operation and high-quality results in your custom printing projects. Even with a deep understanding of basic heat press techniques, it's inevitable to encounter challenges that can affect the outcome of your work. Recognizing and addressing these issues promptly can save time, resources, and ensure customer satisfaction.

One of the most frequent issues encountered involves uneven pressure, which can lead to incomplete or patchy transfers. This problem often stems from an improperly adjusted pressure setting or a warped lower platen. To correct this, ensure that the pressure adjustment knob is set correctly for the material thickness and verify that the platen is flat and undamaged. Regularly checking and calibrating the pressure across the entire platen can prevent these issues from arising.

Another common problem is inaccurate temperature readings, which can affect the quality of the transfer. If the heat press is not reaching the set temperature or if there are fluctuations, it may be due to a malfunctioning thermostat or a faulty heating element. Using an external thermometer to check the accuracy of the machine's temperature display can help diagnose this issue. If discrepancies are found, consulting the manufacturer's guide for troubleshooting steps or seeking professional repair services is advisable.

Adhesion issues, where the design does not fully transfer or peels off after application, are often related to incorrect temperature, time, or pressure settings. Each type of transfer material and fabric has specific requirements that must be meticulously followed. Reviewing the manufacturer's recommendations for each material and ensuring your heat press settings match these specifications is crucial for avoiding adhesion problems.

Scorching or burning of the fabric is a distressing problem that can ruin materials. This usually occurs when the temperature is too high or the item is pressed for too long. To prevent this, always use the lowest possible temperature that still allows for effective transfer and keep pressing times to the minimum recommended duration. Protective silicone pads or Teflon sheets can also help shield sensitive fabrics from direct heat.

Moisture issues can also disrupt the transfer process, leading to bubbles or uneven transfers. Pre-pressing the garment for a few seconds before applying the transfer can eliminate excess moisture and wrinkles, ensuring a smooth and even surface for the transfer.

Ghosting, where the image appears to have a shadow or blur after pressing, is typically caused by the fabric moving during the lifting of the heat press or if the transfer paper shifts. To combat this, ensure the fabric and transfer paper are securely in place before pressing and avoid moving the garment before it has cooled down, if possible.

Finally, addressing mechanical issues such as squeaking joints or a sticky platen requires regular maintenance of the heat press machine. Lubricating moving parts and keeping the platen clean from adhesive residue can prevent many of these problems.

In summary, successfully troubleshooting common issues in basic heat press techniques involves a keen understanding of the machine's operation and the specific requirements of each printing project. By paying close attention to pressure, temperature, and time settings, and by performing regular maintenance, most problems can be avoided or resolved quickly, ensuring the production of high-quality custom apparel and items.

Chapter: 5 Advanced Heat Press Techniques

Layering Techniques for Multi-Colored Designs

Layering techniques for multi-colored designs represent a sophisticated method within the realm of heat press applications, allowing for the creation of vibrant, intricate designs that stand out on any material. This advanced technique leverages the heat press's capabilities to layer different colors and types of heat transfer materials, crafting designs that can range from simple two-tone images to complex, multi-colored graphics. Understanding the nuances of layering is pivotal for anyone looking to elevate their heat press projects to new levels of creativity and professionalism.

The foundation of successful layering lies in the meticulous planning and design phase. Creators must first conceptualize their designs with layering in mind, separating colors and elements that will be applied in sequence. This step is crucial as it determines the order of application, ensuring that colors do not bleed into one another and that the final design aligns perfectly. Software tools that specialize in graphic design play a vital role here, allowing for precise manipulation and preparation of designs before they ever make it to the heat press.

Material selection is another critical aspect of layering. Not all heat transfer materials are suitable for layering, with some types being too thick or not designed to adhere to other layers effectively. Therefore, selecting the right materials that can bond well while retaining their color and integrity under the heat press is essential. Thin, flexible materials are often preferred for layered designs as they offer less bulk and a smoother finish.

The actual process of layering involves careful application of each color or element, one at a time. The first layer is pressed, and then additional layers are added sequentially. Between each layer, it's vital to adjust the heat press settings if necessary, particularly the pressure, to accommodate the increasing thickness of the material. The temperature may also need slight adjustments depending on the materials used, as some may require a lower temperature to prevent burning or discoloration.

One of the more nuanced challenges in layering is managing the heat exposure of each layer. Excessive heat can warp or distort previously applied layers, so it's often recommended to reduce the pressing time for subsequent layers. Some heat transfer materials are designed to be applied for a short duration at a final press to ensure all layers are fully bonded without damaging earlier applications.

Alignment is a key factor in layering that cannot be overstated. Misalignment can throw off the entire design, leading to unsatisfactory results. Using alignment tools or laser guides can help ensure each layer is precisely positioned. Additionally, the use of tacky carriers or repositionable heat transfer materials can offer a margin of error, allowing for minor adjustments before the heat is applied.

To achieve depth and texture in multi-colored designs, some creators experiment with the tactile effects of layering. For example, combining matte and glossy finishes or varying the thickness of layers can add a unique touch to the final product. Such details can transform a simple design into a standout piece, showcasing the creator's skill and the versatility of the heat press.

In conclusion, mastering layering techniques for multi-colored designs requires a blend of careful planning, material knowledge, precise execution, and creativity. It pushes the boundaries of what's possible with a heat press, opening up a world of design possibilities. Whether for business or personal projects, those who invest the time to learn and perfect these techniques can achieve remarkable results, setting their work apart in the competitive landscape of custom apparel and personalized items.

Working with Different Types of Transfers

Mastering the use of a heat press involves understanding the nuances of working with different types of transfers, an area where creativity meets technical expertise. The versatility of a heat press is showcased through its ability to accommodate a wide range of transfer methods, each offering unique benefits and suited for various applications. This advanced technique opens up a world of possibilities for custom apparel and merchandise, demanding a detailed exploration of the main types of transfers: vinyl, inkjet and laser printed transfers, sublimation, and plastisol transfers.

Vinyl transfers stand out for their durability and vibrant colors, making them ideal for bold graphics and text-based designs. Cutting and weeding vinyl require precision, but the result is a professional-grade, wash-resistant finish that appeals to businesses and sports teams. The heat press activates the adhesive on the vinyl, creating a bond with the fabric that withstands numerous washes. Adjusting the heat press settings to suit the specific type of vinyl is crucial, as is ensuring the correct pressure and placement on the material.

Inkjet and laser printed transfers cater to those seeking to reproduce detailed, multi-colored designs without the need for cutting and weeding. Special transfer paper designed for either inkjet or laser printers allows users to print designs at home, which can then be transferred onto fabric using a heat press. This method is particularly appealing for photographs and intricate graphics. However, achieving longevity and quality with these transfers requires understanding the appropriate heat press settings for the type of paper and ink used, as well as the fabric composition.

Sublimation printing is revered for its ability to produce vibrant, full-color images that are integrated into the fabric rather than sitting on top. This technique involves transferring a design printed on special sublimation paper onto polyester or polymer-coated items using a heat press. The high heat converts the solid dye particles into gas, which then bonds with the fabric, resulting in a permanent, fade-resistant image. Sublimation is unparalleled for custom apparel and promotional items that require a soft hand feel and durability. Mastery of sublimation involves not only the correct heat press settings but also an understanding of how colors and designs will interact with different materials.

Plastisol transfers, traditionally associated with screen printing, involve printing a design onto a special release paper using plastisol inks. The heat press then applies the design to the fabric, allowing for the high-quality, textured finishes associated with screen printing without requiring multiple screens for each color. This method is efficient for producing high volumes of prints or for storing designs for future use. The key to success with plastisol transfers lies in the correct application of heat and pressure, ensuring that the ink fully cures and adheres to the fabric.

Navigating the complexities of working with different types of transfers requires not only a deep understanding of the heat press machine but also an appreciation for the characteristics of each transfer type. Factors such as the material of the item being printed, the desired durability of the print, and the complexity of the design all play a role in determining the most suitable transfer method. Advanced users leverage this knowledge to push the boundaries of customization, experimenting with mixed media designs that combine different transfer types for innovative effects.

The journey through advanced heat press techniques is one of experimentation, learning, and creativity. Each type of transfer offers a unique set of challenges and opportunities, encouraging users to explore new ideas and refine their skills. With the support of comprehensive guides and a willingness to experiment, mastering the art of working with different types of transfers becomes not only achievable but also immensely rewarding, opening up endless possibilities in the world of custom apparel and merchandise.

Tips for Sublimation Printing

Sublimation printing stands out as a transformative technique within the realm of heat press applications, enabling the creation of vibrant and lasting designs on a variety of substrates. This advanced method offers unparalleled clarity and color depth, making it a favorite for custom apparel, promotional items, and personalized gifts. To harness the full potential of sublimation printing, understanding and applying a series of refined tips can significantly elevate the quality of the final product.

The foundation of successful sublimation printing lies in the selection of suitable materials. Specifically, synthetic fabrics such as polyester and polymer-coated objects are optimal choices, as they can effectively bond with sublimation dyes at high temperatures. This compatibility is crucial for ensuring that the ink transitions from a solid to a gas and embeds itself into the material, resulting in vivid, durable prints. Experimenting with different substrates can also uncover unique applications and niche market opportunities.

Quality of the initial artwork is paramount in sublimation printing. High-resolution images and designs ensure that the final print maintains clarity and detail. It's recommended to use graphics with a minimum resolution of 300 dpi (dots per inch) and to design in CMYK color mode for color accuracy. Preparing artwork with attention to detail and color will directly influence the vibrancy and appeal of the finished product.

The choice of printer and ink plays a critical role in the sublimation process. Specialized sublimation printers and inks are designed to handle the high heat required to transfer the design onto the substrate. Investing in quality printing supplies ensures consistency, reliability, and efficiency, minimizing issues such as color shifting or fading over time. Regular maintenance of the printer and adherence to manufacturer recommendations can prevent common printing errors and extend the lifespan of the equipment.

Temperature, pressure, and time settings are the trifecta of variables that dictate the success of the sublimation process. These settings can vary depending on the substrate and the specific characteristics of the heat press machine. Conducting test presses is a valuable practice for determining the ideal conditions for each project. This experimentation can help avoid common pitfalls such as ghosting, where the image prints slightly off-target, or blurring, which can occur if the substrate moves during the transfer.

The use of proper accessories and tools, such as heat-resistant tape, Teflon sheets, and silicone pads, enhances the quality and consistency of sublimation prints. These accessories help to distribute heat and pressure evenly while protecting both the substrate and the heat press machine. Additionally, aligning the transfer paper precisely and securing it with heat-resistant tape can prevent shifting and ensure a sharp, clean transfer.

Post-press considerations, including cooling and finishing, are essential steps in the sublimation process. Allowing the printed item to cool properly before handling can prevent smudging or distortion of the image. Understanding how different materials react post-sublimation is important for quality control and can influence decisions on handling, packaging, and presenting the final product.

Lastly, staying informed about the latest trends, materials, and techniques in sublimation printing can inspire new creative ideas and improve efficiency. Engaging with a community of heat press and sublimation enthusiasts through forums, workshops, and trade shows can provide valuable insights and foster innovation.

Incorporating these tips into the sublimation printing process enhances the ability to produce exceptional, high-quality items that stand out in the market. As an advanced technique, sublimation printing requires practice, experimentation, and attention to detail, but mastering this process opens up a world of creative possibilities and business opportunities.

Using Heat Press for Vinyl Applications

Utilizing a heat press for vinyl applications represents a fusion of traditional craftsmanship with modern technology, offering a gateway to producing durable and visually striking designs on a multitude of surfaces. This advanced heat press technique transcends the basic transfer of images onto fabric, delving into the realm of intricate designs and robust applications that withstand the test of time and use.

The essence of using a heat press for vinyl applications lies in its ability to accurately transfer heat-sensitive vinyl onto various substrates. This process requires a nuanced understanding of the relationship between temperature, pressure, and time—variables that are meticulously managed to ensure the vinyl adheres flawlessly without compromising the material or the design itself. The precision afforded by a heat press is unparalleled, ensuring each corner and curve of the vinyl melds seamlessly with the substrate.

Vinyl comes in two primary types: adhesive vinyl and heat transfer vinyl (HTV). Adhesive vinyl is used for stickers and decals, whereas HTV is specifically designed for fabric applications. HTV projects are where the heat press truly shines, allowing for the creation of custom apparel, tote bags, hats, and more with designs that are vibrant, durable, and fully integrated into the fabric.

The process begins with selecting the right type of vinyl for the project. HTV, for instance, is available in a range of finishes from matte to glitter, each adding a unique texture and visual effect to the final product. Once the design is cut and weeded (the process of removing excess vinyl), the heat press comes into play. The correct temperature and pressure settings are crucial; too hot, and the vinyl may warp or the fabric may burn; too low, and the vinyl won't adhere properly. Each type of vinyl and substrate combination has its optimal conditions, often provided by the vinyl manufacturer, which serves as a valuable reference for achieving perfect results.

An advanced technique in using a heat press for vinyl applications is layering. This method involves applying multiple layers of vinyl to create dynamic, multi-colored designs. The challenge lies in doing so without compromising the integrity of the vinyl or the garment. It requires careful planning of the order in which colors are applied, ensuring that the heat press is set to the appropriate settings for each layer. Some vinyl types can be directly layered on top of each other, while others require a cover sheet to protect previously pressed layers. Mastery of layering techniques opens up endless possibilities for custom designs and finishes.

Another sophisticated aspect of using a heat press for vinyl applications is working with stretchable HTV on elastic fabrics. This requires a special kind of vinyl that can stretch with the fabric without cracking or peeling. The heat press settings, in this case, need to be adjusted to accommodate the delicate balance between adhesion and flexibility.

The heat press also excels in precision placement, a critical factor when working with small or intricate designs. Accessories like laser alignment tools and placement guides help ensure that the vinyl is positioned accurately before pressing. This precision is particularly important for creating professional-quality apparel and items where the placement of the design contributes significantly to the overall aesthetic.

Moreover, the efficiency and speed of a heat press make it ideal for small businesses or individuals looking to produce high-quality custom merchandise on demand. Compared to other methods of vinyl application, the heat press significantly reduces production time while maintaining or improving the quality of the finished product.

In conclusion, using a heat press for vinyl applications is a skill that combines art with science. It demands a comprehensive understanding of the materials involved and a meticulous approach to the technical aspects of the heat press machine. For those willing to delve into the complexities of this technique, the rewards are substantial—enabling the creation of personalized, durable, and professional-grade products that stand out in the market. Whether for business or personal projects, mastering this advanced heat press technique is a valuable addition to any creator's toolkit.

Chapter: 6 Maintenance and Troubleshooting

Daily and Periodic Maintenance Tips

Maintaining a heat press to ensure its longevity and optimal performance involves a blend of daily routines and periodic checks that collectively prevent common issues and ensure the machine operates smoothly. Understanding these maintenance practices is crucial for anyone using a heat press, as it helps avoid downtime, extends the life of the equipment, and ensures consistent quality in the production of custom apparel and other personalized items.

Starting with daily maintenance, the primary focus is on keeping the machine clean and free from residues that can accumulate from the transfer process. After each use, it's essential to wipe down the platen, or heating element, with a soft, dry cloth to remove any leftover ink, adhesive, or other substances that might have been transferred during the pressing process. For tougher residues, using a mild cleaning agent designed for heat press machines, applied to a cloth rather than directly on the platen, can effectively remove buildup without damaging the surface. It's important to ensure the machine is cooled down before attempting any cleaning to avoid burns or other injuries.

The silicone pad and bottom table of the heat press also require attention. These should be inspected daily for any signs of wear, debris, or adhesive buildup and cleaned accordingly. A well-maintained pad ensures even pressure distribution during the pressing process, which is vital for the quality of the final product.

Moving on to periodic maintenance, this involves more in-depth checks and tasks that may not need to be performed daily but are crucial for the overall health of the machine. These checks include inspecting the electrical connections and cables for signs of wear or damage, ensuring that all components are functioning correctly and safely. Loose connections can lead to inconsistent heating or operational failures, so tightening and securing all connections is a preventive measure against such issues.

Another critical aspect of periodic maintenance is checking the pressure adjustment mechanism. Over time, the mechanism can become stiff or lose its accuracy, affecting the quality of the press. Lubricating moving parts with a high-temperature lubricant can help maintain smooth operation and prevent wear. Additionally, recalibrating the pressure settings may be necessary to ensure that the machine continues to apply the correct amount of pressure for different materials.

Temperature accuracy is another area that requires periodic attention. Using a separate temperature gauge, verify that the heat press is reaching and maintaining the correct temperatures as displayed. Inaccuracies can lead to improperly cured transfers, affecting the durability and quality of the print. If discrepancies are found, consulting the manufacturer's guidelines or a professional service may be necessary to recalibrate or repair the heating element.

The overall structure and alignment of the heat press should also be inspected periodically. Misalignment can lead to uneven pressure and heat distribution, which are critical for achieving high-quality prints. Checking the alignment of the upper and lower platens and making adjustments according to the manufacturer's instructions can prevent these issues.

Finally, keeping a log of maintenance activities, including both daily and periodic tasks, can help track the machine's condition and identify when service or repairs might be needed. This proactive approach to maintenance not only extends the life of the heat press but also ensures that it operates efficiently and produces high-quality results consistently.

By adhering to these daily and periodic maintenance tips, users can significantly reduce the risk of operational issues, maintain the quality of their printed products, and maximize the lifespan of their heat press machine. This routine care is an investment in the reliability and productivity of one's printing capabilities, ensuring that the heat press remains a valuable asset in the creation of custom apparel and personalized items.

Troubleshooting Common Problems

In the journey of mastering the use of a heat press, encountering problems can be both a challenge and an opportunity for growth. By understanding common issues and learning how to troubleshoot them effectively, users can maintain their equipment in optimal condition and ensure the quality of their printed products remains high.

One of the most frequent issues faced is uneven pressure, which can lead to incomplete or patchy transfers. This problem often stems from an improperly aligned platen or an uneven surface. To resolve this, users should check the heat press for any visible misalignment and adjust the pressure settings according to the manufacturer's instructions. Regularly inspecting the silicone pad for wear and tear is also crucial, as a degraded pad can affect pressure distribution.

Another common challenge is inaccurate temperature readings, which can result in transfers that are either under or over-cured. This discrepancy is often due to a malfunctioning thermostat or a build-up of residue on the heating element, which affects its efficiency. Regular cleaning of the heating element and calibration of the temperature gauge can help ensure accurate readings. If the problem persists, it may be necessary to replace the thermostat.

The appearance of scorch marks or heat stains on fabric is another issue that can compromise the quality of the final product. This problem is typically caused by setting the temperature too high or leaving the press closed for too long. To prevent this, it's important to adhere to the recommended temperature and time settings for the specific material and transfer type being used. Testing the settings on a scrap piece of material before proceeding with the actual project can also help avoid this issue.

Adhesive residue on the platen or garments can cause transfers to stick or peel prematurely. This sticky situation is often the result of using too much adhesive or not using a protective cover sheet during the transfer process. Cleaning the platen regularly with an appropriate solvent and always using a protective sheet can mitigate this issue.

Electrical problems, such as the heat press not heating up or turning on, can be particularly frustrating. These issues may be due to a faulty electrical outlet, a damaged power cord, or internal wiring issues. Checking the power source and inspecting the cord for damage are initial steps to troubleshoot this problem. If these components appear to be in good working order, consulting a professional technician or the manufacturer's support team is advisable.

Occasionally, users may encounter difficulty with the digital display, such as unresponsive buttons or inaccurate timer settings.

This problem can often be resolved by resetting the heat press according to the manufacturer's instructions. If the display continues to malfunction, it may be a sign of a deeper electrical issue that requires professional attention.

Regular maintenance is the key to preventing many common problems with heat presses. This includes keeping the machine clean, checking for loose or worn parts, and following the manufacturer's guidelines for use and care. By adopting a proactive approach to maintenance and troubleshooting, users can extend the life of their heat press and maintain a high standard of quality in their printed products.

In conclusion, while encountering problems with a heat press can be daunting, understanding how to effectively troubleshoot these issues is an essential skill. By addressing problems promptly and conducting regular maintenance, users can ensure their heat press operates efficiently, allowing them to continue producing high-quality custom apparel and items with confidence.

When to Seek Professional Help

When navigating the intricacies of operating a heat press, both newcomers and seasoned users can encounter situations that necessitate professional intervention. Understanding when to seek professional help is critical to maintaining the longevity and performance of your heat press, ensuring that it continues to produce high-quality prints without compromising safety.

A clear indicator for seeking professional help is when you notice inconsistent heat distribution across the platen. This can lead to uneven transfers and significantly impact the quality of your finished products. Such issues often stem from malfunctioning heating elements or faulty wiring, problems that require technical expertise to diagnose and repair safely.

Similarly, if the heat press exhibits inaccurate temperature readings or fails to reach the set temperature, it's time to consult with a professional. These symptoms could be a sign of a malfunctioning thermostat or temperature control system. Attempting to fix these issues without proper knowledge can lead to further damage to the machine or pose a risk of injury.

Mechanical issues, such as difficulty in opening or closing the press or an unresponsive pressure adjustment mechanism, also warrant professional attention. These problems could indicate wear and tear on the machine's physical components, which, if left unaddressed, can exacerbate over time and lead to a complete breakdown.

Electrical issues are particularly concerning due to the potential risk of electric shock or fire. Signs of electrical problems include the heat press tripping circuit breakers, emitting a burning smell, or displaying error messages related to the electrical system. In such cases, immediately disconnect the machine from power and contact a professional repair service.

Moreover, when you encounter persistent problems that defy basic troubleshooting steps outlined in the machine's manual, seeking professional help can save time and resources. Professionals can offer a comprehensive diagnosis that identifies underlying issues which may not be immediately apparent to the user.

For those who rely on their heat press for business, downtime can translate into lost revenue and dissatisfied customers. Professional technicians can often provide quicker and more efficient repairs than attempting to troubleshoot issues on your own, minimizing the impact on your operations.

Lastly, it's advisable to seek professional maintenance services periodically, even in the absence of obvious problems. Regular professional maintenance can prevent many common issues from developing and extend the lifespan of your heat press. Technicians can perform a thorough inspection, clean parts that are difficult to reach, and replace worn components before they fail.

In conclusion, while many minor issues with a heat press can be resolved through routine maintenance and troubleshooting, certain situations require the expertise of a professional. By recognizing the signs that indicate a need for professional help, you can ensure the safety, efficiency, and longevity of your heat press, maintaining its role as a cornerstone in your creative endeavors or business operations.

Chapter: 7 Creative Projects and Ideas

Custom T-Shirts

Custom t-shirts stand as a cornerstone of personal expression and business branding, offering a canvas that's as versatile as it is visible. The advent of heat press technology has revolutionized the way these t-shirts are created, making it possible for anyone from hobbyists to professional print shops to produce high-quality, vibrant designs with precision and efficiency.

At the heart of custom t-shirt creation with a heat press is the ability to transfer intricate designs, photographs, and text onto fabric, resulting in a product that can withstand the test of time and frequent washing without fading or peeling. This durability is a key advantage, ensuring that the effort and creativity invested in each design have a long-lasting impact. The heat press technique also allows for the reproduction of full-color images with impressive detail, opening up endless possibilities for creativity and customization.

The process begins with the design phase, where digital artworks or photographs are prepared using graphic design software. This digital design is then printed onto special heat transfer paper. The heat press is prepped by setting the correct temperature, pressure, and time for the specific material of the t-shirt, ensuring that the transfer is executed flawlessly. Once the press is heated to the desired temperature, the t-shirt is placed on the machine, the transfer paper is positioned on top, and the press is activated. In just a few moments, the design is permanently transferred to the t-shirt, ready to be worn, sold, or given as a personalized gift.

One of the most compelling aspects of creating custom t-shirts with a heat press is the flexibility it offers. Whether producing a one-off design for a special occasion or mass-producing merchandise for a business, the heat press can accommodate various production scales without compromising quality. This flexibility is particularly valuable for small businesses and startups, as it allows them to offer customized products without the need for large inventory stocks.

Moreover, custom t-shirts are not just about the visuals; they're a powerful tool for communication. They can be used to promote brands, support causes, commemorate events, or simply express personal style. The accessibility of heat press technology means that creating a custom t-shirt is no longer a costly endeavor reserved for professionals. With some practice and creativity, anyone can bring their t-shirt ideas to life.

The environmental aspect is also worth noting. The heat press process is relatively clean, with minimal waste compared to traditional screen printing, making it a more sustainable choice for eco-conscious creators and consumers.

In conclusion, the creation of custom t-shirts using a heat press embodies the intersection of art, technology, and fashion. It democratizes the process of garment printing, offering a high-quality, efficient, and versatile method of production. Whether for business or personal use, custom t-shirts serve as a dynamic form of expression, and learning to master the heat press opens up a world of potential for creators eager to make their mark on the fabric of society.

Personalized Bags and Accessories

The world of personalized bags and accessories offers a canvas for creativity and personal expression that resonates with a wide audience, from fashion enthusiasts to gift seekers. Through the use of a heat press, individuals and businesses alike unlock the potential to transform ordinary items into unique, customized treasures. This technique not only provides a platform for personal flair but also serves as a lucrative avenue for those looking to venture into the realm of custom merchandise.

Personalizing bags and accessories with a heat press involves a process that imbues these items with a level of quality and uniqueness that off-the-shelf products seldom achieve. Whether it's totes, clutches, backpacks, caps, or wallets, each piece becomes a testament to one's style or a tailored message intended for a special someone. The appeal of customized items lies in their ability to stand out, making them not just accessories but statements of individuality.

The process begins with the selection of designs, which can range from simple logos and monograms to intricate artworks and photographs. This diversity allows for a broad spectrum of styles, catering to different tastes and occasions. Heat press technology ensures that these designs are transferred onto fabric or other materials with precision, resulting in vibrant colors and sharp details that endure wear and tear. Unlike other printing methods, the heat press technique secures the ink deep within the fibers, enhancing the durability of the design and maintaining its appearance over time.

One of the key advantages of using a heat press for creating personalized bags and accessories is the ability to produce items on demand. This flexibility means that products can be customized in small batches or even as one-off pieces, reducing waste and avoiding the need for large inventory spaces. For businesses, this approach enables the offering of bespoke services, where customers can have their items customized to their exact specifications, adding a personal touch that enhances the value of the product.

Furthermore, the speed and efficiency of the heat press process allow for quick turnaround times, making it possible to cater to last-minute orders or rapid production needs. This efficiency is particularly beneficial during peak seasons or special events, where the demand for personalized gifts and accessories surges.

Innovative uses of heat press technology in personalizing bags and accessories have also opened up new creative possibilities. For instance, the use of specialty inks and materials, such as glow-in-the-dark or reflective transfers, adds a unique dimension to products. Additionally, the ability to work with a variety of materials means that customization can extend beyond fabric, to include leather, synthetic blends, and more, broadening the scope of products that can be offered.

The environmental aspect is another consideration that adds to the appeal of using a heat press. The process is cleaner, generating less waste compared to traditional printing methods, and does not require water or emit harmful substances. This eco-friendly approach not only aligns with the growing consumer preference for sustainable products but also contributes to the responsible operation of businesses.

In conclusion, the realm of personalized bags and accessories presents a fertile ground for creative expression and business opportunities, facilitated by the use of heat press technology. The ability to produce high-quality, durable, and uniquely customized items on demand offers a competitive edge in the market, appealing to consumers looking for products that reflect their personality or convey a special message. As this industry continues to evolve, the heat press remains a vital tool for unlocking the full potential of personalization, merging creativity with innovation to create products that resonate on a personal level.

Home Decor Items

The versatility of a heat press extends far beyond the realm of custom apparel, diving into the transformative world of home decor. By harnessing the power of heat press technology, creators can infuse personal flair and uniqueness into various home items, making custom home decor not just a possibility but an exciting avenue for artistic expression and business ventures.

One of the most compelling aspects of using a heat press for home decor is the ability to personalize items with designs, patterns, and images that reflect individual tastes or current trends. This personalization is not just limited to aesthetics; it allows for the creation of meaningful gifts and unique pieces that carry sentimental value, whether it's through customized cushions that feature beloved family photos or kitchen towels adorned with witty quotes.

Cushion covers stand out as a popular choice for heat press projects. They serve as blank canvases ready to be transformed by vibrant colors and intricate designs. The heat press ensures that these designs are not only visually stunning but also durable enough to withstand the rigors of daily use, making them both practical and decorative.

Wall art, another domain where the heat press makes a significant impact, allows for the creation of custom canvas prints. Whether it's a reproduction of a classic masterpiece or a modern graphic design, the heat press can transfer these images onto canvas, giving any room an instant lift. This method provides an affordable way to fill homes with beautiful art that can easily be swapped out or updated as decor tastes evolve.

Personalized coasters and placemats bring a touch of individuality to the dining table. With a heat press, these items can be customized to match the theme of a dinner party, celebrate a special occasion, or simply complement the home's decor. The ability to use heat-resistant materials means these items are not just decorative but fully functional as well.

For those looking to add a cozy touch to their space, custom-made throw blankets created with a heat press offer endless possibilities. From family heirlooms featuring generational photos to modern designs that match the living room palette, these blankets combine comfort with personalization.

The bathroom need not be left out, with the possibility of creating custom shower curtains that turn a typically overlooked space into a showcase of creativity. Whether it's a tranquil landscape to create a spa-like atmosphere or a bold geometric pattern for a modern twist, the heat press brings these visions to life.

In addition to these items, the heat press can be used to create door mats, window treatments, and even lampshades, each offering a unique opportunity to inject personality and style into home decor. The process not only elevates the aesthetic appeal of these items but also ensures that the designs are long-lasting and can stand up to the demands of daily use.

Using a heat press for home decor items opens up a world of creative possibilities, allowing individuals and businesses alike to offer customized and unique products. It bridges the gap between personal expression and functional design, making it an invaluable tool for anyone looking to explore the potential of custom home decor. Whether for personal use, gifting, or commercial ventures, the heat press stands as a gateway to endless creativity in the home decor space.

Unique Gift Ideas

In a world where personalization is highly valued, the heat press stands out as a versatile tool capable of turning ordinary items into extraordinary, unique gifts. The ability to transfer custom designs onto various materials allows for the creation of personalized presents that resonate on a personal level, making them more meaningful and memorable. Here are some creative projects and ideas utilizing a heat press that can inspire those looking to craft one-of-a-kind gifts.

Custom T-Shirts are a classic choice, offering a blank canvas to express creativity. Whether it's a funny quote, a favorite image, or a design that captures a special moment, custom T-shirts make for a thoughtful gift that's both personal and practical. With a heat press, creating these customized gifts becomes an effortless process, allowing for the reproduction of high-quality, vibrant designs that last.

Personalized Bags, such as tote bags, backpacks, and cosmetic bags, serve as excellent gifts for friends and family. Using a heat press to apply unique artwork, monograms, or meaningful messages can transform these everyday items into cherished personal accessories. They're perfect for birthdays, weddings, or as thoughtful bridesmaids' gifts, providing a stylish and useful reminder of a special occasion.

Customized Home Decor items like throw pillows, blankets, and wall hangings allow for the incorporation of personal touches into home spaces. A heat press can be used to apply personalized designs, photos, or quotes that add a warm, personal vibe to any room. These items not only serve as decorative pieces but also as cozy reminders of the thoughtfulness of the giver.

Personalized Drinkware, such as mugs and water bottles, can also be customized using heat press technology. By transferring photos, unique designs, or witty quotes onto these everyday items, they become much more than just drinkware; they become a daily reminder of a special relationship or an inside joke shared among friends.

DIY Patches and Appliques created with a heat press can be applied to a wide range of items, including jackets, jeans, hats, and backpacks, allowing for a high level of customization. These small but impactful additions can transform wardrobe staples into statement pieces that reflect the recipient's personality and interests.

Photo Panels and Canvas Prints are another unique gift idea made possible with a heat press. High-quality photos can be transferred onto canvas or other materials to create beautiful, durable artwork that captures life's precious moments. These gifts are perfect for anniversaries, graduations, and as a way to preserve memories in a tangible form.

Custom Pet Accessories, like bandanas or small blankets, are perfect for pet owners. A heat press can be used to apply cute designs, pet names, or even photo transfers, making for a gift that both pets and their owners will adore.

The use of a heat press in creating unique gifts opens up a world of possibilities for personalized gifting. The ability to customize items with personal touches not only enhances the value of the gift but also strengthens the bond between the giver and the recipient. Through creative projects and ideas, the heat press becomes a tool for expressing love, appreciation, and thoughtfulness in a way that mass-produced items simply cannot match.

Chapter: 8 Expanding Your Business with Heat Pressing

Marketing Your Products

In the dynamic world of custom apparel and merchandise, mastering the art of heat pressing is just the beginning. The true challenge—and opportunity—lies in effectively marketing your products to reach a wider audience and expand your business. As you navigate the complexities of promotion and sales in relation to heat pressing, several strategic approaches can significantly enhance your visibility and appeal to potential customers.

Developing a strong brand identity is foundational to marketing success. Your brand should resonate with your target audience, reflecting the quality, creativity, and uniqueness of the items you produce with your heat press. This includes a memorable logo, a consistent color scheme, and a voice that communicates your brand's values and mission across all platforms. By establishing a distinct brand, you set yourself apart in a crowded market, making your products instantly recognizable and more desirable to consumers.

In today's digital age, an online presence is indispensable. Creating a user-friendly website showcasing your product range allows customers to browse and purchase your offerings with ease. Integrating high-quality images and detailed descriptions of each item not only highlights the craftsmanship but also educates potential buyers on the value of heat-pressed goods. Additionally, leveraging social media platforms can amplify your reach, enabling you to connect with a global audience, share behind-the-scenes content, and promote new products or special offers.

Content marketing, through blogs, videos, and tutorials, can further engage your audience by providing valuable information related to your niche. For example, sharing insights into the heat pressing process, design tips, or the latest trends in custom apparel not only positions you as an authority in your field but also builds trust with your audience. This approach not only attracts interested viewers but also encourages them to share your content, expanding your reach organically.

Email marketing remains a powerful tool for direct communication with your customers. By collecting email addresses through your website or at events, you can create targeted campaigns that inform subscribers of new products, upcoming sales, or exclusive deals. Personalizing these communications increases engagement and loyalty, turning one-time buyers into repeat customers.

Participating in local markets, trade shows, and community events can also play a crucial role in marketing your products. These venues offer the opportunity to showcase the quality and versatility of your heat-pressed items in person, allowing potential customers to see, touch, and appreciate the craftsmanship firsthand. Networking with other vendors and attendees can lead to valuable partnerships, expanding your visibility and customer base.

Collaborations with influencers or local organizations can boost your marketing efforts by introducing your products to new audiences. Partnering with individuals or groups that align with your brand values and have a strong following in your target market can be particularly effective. Whether through sponsored posts, co-branded merchandise, or event sponsorships, these collaborations can significantly increase your brand's exposure and credibility.

Finally, offering exceptional customer service is a critical aspect of marketing. Satisfied customers are more likely to return and recommend your business to others. Encourage feedback, address concerns promptly, and go above and beyond to meet your customers' needs. Positive reviews and word-of-mouth referrals can be incredibly valuable in attracting new business.

In summary, marketing your heat-pressed products effectively requires a multifaceted approach that encompasses brand development, digital presence, content creation, direct engagement, community participation, strategic partnerships, and outstanding customer service. By employing these strategies, you can not only reach a wider audience but also build a loyal customer base that supports your business's growth and success.

Pricing Strategies

Expanding your business with heat pressing involves not just mastering the technical aspects of the equipment but also developing effective pricing strategies. As you venture into the realm of custom apparel and merchandise, setting the right prices is crucial for attracting customers, maximizing profits, and ensuring the long-term success of your venture. The challenge lies in finding a balance between competitive pricing and covering costs, all while delivering value to your customers.

One of the first steps in establishing a pricing strategy is understanding your costs. This includes both the direct costs associated with producing each item, such as materials, ink, and wear on your heat press machine, and the indirect costs, like utilities, rent for your workspace, and the time you invest in designing and creating products. Calculating these costs accurately provides a solid foundation for setting prices that ensure profitability.

Market research plays a pivotal role in shaping your pricing strategy. By analyzing competitors' prices, you can gauge the going rate for similar products in your niche. This insight helps you position your offerings in a way that is competitive yet still reflects the quality and uniqueness of your products. Remember, while it's important to stay competitive, underpricing can undervalue your work and make it difficult to cover costs or expand your business.

Value-based pricing is another strategy that can be particularly effective for businesses utilizing heat pressing. This approach involves setting prices based on the perceived value of your products to the customer rather than solely on your costs or the market average. For example, custom designs or limited-edition items may carry a higher value and thus command higher prices. This strategy requires a deep understanding of your target market and what aspects of your products they value most, whether it's the quality, customization options, or the exclusivity of your designs.

Dynamic pricing is a flexible strategy that allows you to adjust prices based on various factors, such as demand, seasonal trends, or the introduction of new products. For instance, you might increase prices during peak shopping seasons or when launching a highly anticipated product line. Conversely, lowering prices for clearance items or older stock can help move inventory and make room for new designs. This approach requires careful monitoring of sales and market trends but can maximize profits and customer interest over time.

Bundling products is a strategy that can both provide value to your customers and increase your average order value. By offering related items at a discounted rate when purchased together, you encourage customers to buy more while perceiving a greater value. For instance, bundling a custom T-shirt with a matching tote bag or hat creates an attractive offer that can boost sales and deepen customer engagement with your brand.

Lastly, transparency in pricing is vital for building trust with your customers. Clearly communicating the costs of customization, shipping, and any other fees helps avoid surprises and ensures a positive buying experience. Providing a breakdown or explaining the factors that contribute to the price of your products can also highlight the quality and care that go into each item, justifying the price and reinforcing the value proposition to your customers.

In summary, developing a comprehensive pricing strategy is a critical component of expanding your business with heat pressing. By understanding your costs, conducting market research, applying value-based and dynamic pricing, offering bundled products, and maintaining transparency, you can set prices that attract and retain customers, cover costs, and maximize profitability. As you grow your business, regularly reviewing and adjusting your pricing strategy will ensure it remains aligned with your goals, market conditions, and customer expectations.

Scaling Production

Scaling production in the context of expanding a business with heat pressing involves a strategic approach to increasing output without compromising on quality or efficiency. This expansion is not just about meeting increased demand but also about positioning the business for sustainable growth and competitive advantage. As entrepreneurs delve deeper into the world of custom apparel and personalized merchandise using heat presses, understanding how to effectively scale production becomes paramount.

At the foundation of scaling production is the optimization of existing processes. This means revisiting the workflow to identify bottlenecks and inefficiencies that could hinder scaling efforts. For businesses using heat presses, this could involve streamlining the design preparation, transfer process, and post-press activities. Efficient workflow management ensures that production can ramp up smoothly, minimizing wasted time and resources.

Investment in additional or more advanced heat press equipment is often a necessary step in scaling production. Upgrading from a basic heat press to more sophisticated models with larger platens, multiple stations, or automated features can significantly increase output. Such machines can handle larger volumes and sometimes even operate with less direct manual intervention, freeing up time and resources to focus on other aspects of the business.

Training and expanding the team is another critical component. As production scales, the need for a skilled workforce to operate heat press machinery, manage quality control, and handle additional orders becomes more apparent. Investing in training for current employees and hiring new team members with the necessary skills or potential for development supports a growing operation. A well-trained team not only increases production capacity but also maintains the quality that customers expect.

Implementing technology and software solutions can further streamline the production process. From graphic design software that optimizes designs for heat pressing to inventory and order management systems that keep track of materials and customer orders, technology plays a pivotal role in scaling. Automation in certain areas, such as scheduling and logistics, can also enhance efficiency, allowing the business to handle a higher volume of orders with greater accuracy.

Diversifying the product range to include a wider variety of items that can be customized using heat press technology can attract a broader customer base and open up new markets. However, this diversification should be strategic, ensuring that the business can maintain quality and efficiency across all products. Researching market trends and customer preferences can guide decisions on which new products to introduce.

Quality control is essential as production scales. Maintaining high standards while increasing output requires robust quality assurance processes. This might involve setting clearer guidelines for quality, regular training updates for staff on quality expectations, and investing in equipment that provides consistent results. Keeping quality at the forefront ensures customer satisfaction and reduces the costs associated with reworks and returns.

Finally, marketing and customer engagement become increasingly important as a business scales. Effective marketing strategies not only help to sell the increased volume of products but also build brand loyalty and attract new customers. Utilizing social media, email marketing, and other digital platforms can amplify a brand's reach, while engaging with customers through personalized products and services encourages repeat business.

Scaling production with heat pressing requires a balanced approach that encompasses equipment upgrades, process optimization, workforce expansion, technology integration, product diversification, quality control, and strategic marketing. By carefully planning and implementing these strategies, businesses can successfully expand their operations, meet the growing demand for customized merchandise, and achieve sustainable growth.

Exploring New Markets

Expanding your business with heat pressing involves venturing into new markets, a strategic move that can significantly enhance your brand's visibility and profitability. The versatility of heat press technology opens up numerous opportunities for customization and creativity, making it a potent tool for entrepreneurs looking to broaden their horizons.

The first step in exploring new markets is understanding the vast potential of heat pressing. From fashion apparel to promotional merchandise and home decor, the range of products that can be customized is virtually limitless. This diversity allows businesses to target various consumer segments based on interests, demographics, and trends. For instance, a company traditionally focused on custom t-shirts can easily expand into producing personalized tote bags, aprons, or cushion covers, each appealing to different customer bases.

Researching and identifying emerging trends is critical in this expansion process. Staying informed about popular designs, themes, and products can position your business to quickly adapt and cater to new demands. For example, tapping into the eco-friendly products market by offering sustainable and organic fabric options for custom printing could attract environmentally conscious consumers.

Understanding the needs and preferences of your target market is essential. This may involve conducting market research, analyzing customer feedback, and monitoring social media trends to pinpoint what appeals to potential new customers. Tailoring your product offerings to meet these needs not only enhances customer satisfaction but also sets your business apart from competitors.

Leveraging online platforms is another key strategy for exploring new markets. An effective online presence, through a well-designed website and active social media channels, can help you reach a global audience. E-commerce platforms offer a convenient way for customers to browse and purchase customized products, significantly widening your market reach. Additionally, utilizing online advertising and digital marketing strategies can increase visibility and attract customers from previously untapped markets.

Networking and partnerships can also play a vital role in market expansion. Collaborating with influencers, brands, or local businesses can introduce your products to new audiences. For example, offering branded merchandise for events or corporate gifts can open doors to the B2B market, diversifying your customer base.

Innovation is at the core of exploring new markets. Continuously experimenting with new materials, designs, and products can lead to unique offerings that capture attention. For instance, incorporating heat press techniques with other crafting methods, like embroidery or screen printing, can result in one-of-a-kind products that appeal to niche markets.

Customer service excellence remains a cornerstone of successful market expansion. Ensuring a positive customer experience, from product quality to delivery and after-sales support, builds trust and loyalty, encouraging word-of-mouth referrals and repeat business.

Lastly, understanding the legal and regulatory aspects of entering new markets is crucial. This includes compliance with local laws, understanding import/export regulations, and protecting your designs and intellectual property. Being well-informed can prevent costly legal issues and ensure smooth business operations.

In conclusion, exploring new markets by expanding your business with heat pressing involves a combination of creativity, market research, strategic planning, and customer-focused innovation. By leveraging the versatility of heat press technology and adopting a proactive approach to market expansion, businesses can uncover new opportunities for growth and success in an ever-evolving marketplace.

Chapter: 9 Health and Safety Considerations

Operating Safely

Operating a heat press involves more than just creating visually appealing designs; it necessitates a keen awareness of health and safety considerations to prevent accidents and ensure a safe working environment. The high temperatures and pressure required to transfer designs onto materials can pose risks if proper precautions are not taken. Therefore, understanding and implementing safety measures is crucial for anyone using a heat press.

Firstly, personal protective equipment (PPE) is essential. Heat-resistant gloves are a must to protect hands from burns when handling the press or materials immediately after pressing. Safety glasses or goggles are also advisable to guard against the potential risk of materials or parts breaking under pressure and causing eye injuries. Additionally, wearing aprons or other protective clothing can help shield against heat and any hot materials.

Ventilation is another critical factor when operating a heat press, especially when working with materials or transfer agents that may release fumes at high temperatures. A well-ventilated workspace helps to disperse these fumes, reducing the inhalation risk. For enclosed spaces, consider using air filtration systems or exhaust fans to maintain air quality and prevent the buildup of potentially harmful vapors.

Understanding the equipment is key to safe operation. Users should be thoroughly familiar with their heat press machine, including its specific features and safety mechanisms. This includes knowing how to properly adjust pressure and temperature settings, as well as how to use timers to prevent overheating. Reading and adhering to the manufacturer's guidelines can prevent misuse that might lead to accidents or damage to the machine.

Regular maintenance and inspection of the heat press are essential to ensure it remains in good working condition. This includes checking for worn or damaged parts, ensuring electrical cords and plugs are intact and free from damage, and cleaning the machine according to the manufacturer's instructions to prevent the accumulation of lint, dust, or ink residues that could pose a fire hazard.

Emergency preparedness should not be overlooked. Having a fire extinguisher rated for electrical fires within easy reach and knowing how to use it is vital. Similarly, keeping a first aid kit nearby and being familiar with basic first aid procedures for burns and other potential injuries can make a significant difference in response times during an accident.

Furthermore, proper training cannot be understated. Whether it's through formal training sessions, online tutorials, or manufacturer's instructions, understanding how to operate the heat press safely is paramount. This includes recognizing the signs of equipment malfunctioning and knowing when to stop operation and seek repairs.

Lastly, ergonomics plays a crucial role in operating a heat press safely. The repetitive nature of loading and unloading the press, along with standing for extended periods, can lead to strain and fatigue. To mitigate these issues, ensure the work area is set up to allow for natural movements and the heat press is at a comfortable working height. Anti-fatigue mats can also provide cushioning and support for those standing for long periods.

In sum, operating safely with a heat press encompasses a broad range of considerations, from personal protective equipment and ventilation to equipment familiarity and ergonomics. By adhering to these guidelines, users can minimize risks and enjoy a productive, safe working environment. This commitment to safety not only protects the individual but also contributes to the overall success and sustainability of their printing endeavors.

Ventilation and Air Quality

The operation of a heat press, integral to the creation of custom apparel and personalized items, necessitates careful attention to ventilation and air quality to ensure a safe working environment. This focus on health and safety stems from the recognition that the heat pressing process can release various chemicals and fumes, especially when working with certain materials and transfer agents. Maintaining optimal air quality is not only essential for the well-being of those operating the heat press but also for anyone within the vicinity of the workspace.

Ventilation is a critical factor in mitigating the potential hazards associated with these emissions. A well-ventilated area dilutes and removes airborne contaminants, thereby reducing the risk of respiratory issues and other health problems. Effective ventilation strategies include the use of mechanical ventilation systems, such as exhaust fans and air filtration units, which actively draw contaminated air out of the workspace and replace it with fresh air. For smaller operations or those in confined spaces, ensuring that doors and windows are open can also aid in promoting airflow, though this may be less effective in controlling the concentration of contaminants compared to mechanical systems.

The types of fumes and chemicals released during heat pressing vary depending on the materials and transfer methods used. For example, plastisol transfers commonly used in screen printing can emit volatile organic compounds (VOCs) when heated, while sublimation printing involves the transition of solid dyes to gas, potentially releasing particles and fumes that could be inhaled. Additionally, the application of certain adhesives and pre-treatment chemicals may contribute to the overall chemical load in the air.

To further safeguard air quality, operators should consider the use of personal protective equipment (PPE) such as masks or respirators designed to filter out particulates and fumes. This is particularly important in environments where ventilation improvements are not feasible or where high concentrations of contaminants are expected. Regular monitoring of air quality through the use of detectors or professional assessments can also provide valuable insights into the effectiveness of ventilation measures and the need for additional safety protocols.

Educating those who use heat presses about the importance of air quality and the potential risks associated with inadequate ventilation is another vital component of a comprehensive health and safety plan. This includes training on the correct use and maintenance of ventilation equipment, the proper handling and storage of materials that may emit hazardous fumes, and the importance of regular breaks to minimize exposure.

In addition to immediate health concerns, long-term exposure to poor air quality can have significant implications for respiratory health, including the development of chronic conditions such as asthma or other pulmonary diseases. Therefore, it is in the interest of both individuals and businesses to invest in effective ventilation solutions and to foster a culture of safety that prioritizes the well-being of all who interact with heat press technology.

In summary, the relationship between the use of heat presses and the need for vigilant attention to ventilation and air quality is a critical aspect of workplace health and safety. Through the implementation of robust ventilation systems, the use of PPE, regular air quality monitoring, and ongoing education and training, operators can significantly reduce health risks and create a safer, more productive environment for heat pressing activities.

Handling Chemicals and Supplies

When it comes to using a heat press, the focus often lies on the creative possibilities and technical aspects of the process. However, an equally important factor that should never be overlooked is the handling of chemicals and supplies, especially given the health and safety considerations involved. Proper management of these elements is critical, not only for the safety of the operator but also for the longevity of the equipment and the quality of the finished product.

Heat press operations may involve various chemicals, such as dyes, transfer papers, and cleaning agents. These substances can pose risks if not handled correctly. For instance, certain dyes and inks used in the transfer process can emit fumes when heated, some of which may be harmful if inhaled over time. Similarly, cleaning agents used to maintain the heat press can contain volatile organic compounds (VOCs) or other hazardous materials that require careful handling to avoid skin irritation, respiratory issues, or worse.

The first step in safely handling chemicals and supplies is to familiarize oneself with the Material Safety Data Sheets (MSDS) for each substance used. These sheets provide vital information on the properties of the chemicals, including their potential health hazards and instructions for safe handling and first aid measures in case of accidental exposure. Ensuring that all users of the heat press have access to and understand this information is crucial.

Proper ventilation is a key aspect of maintaining a safe environment when operating a heat press. Adequate airflow helps to disperse fumes that may be released during the pressing process, reducing the risk of inhaling harmful substances. Depending on the setup and the chemicals in use, this may require the installation of specialized exhaust systems or simply ensuring that the work area is well-ventilated with fresh air.

Personal protective equipment (PPE) also plays a vital role in safeguarding the operator. Gloves, masks, and eye protection should be considered essential when handling chemicals that can irritate the skin or eyes or when working in an environment where fumes may be present. The specific type of PPE required can vary based on the chemicals being used, so it's important to refer to the MSDS for guidance.

In addition to personal safety, proper storage of chemicals and supplies is necessary to prevent accidents and ensure their longevity. Chemicals should be stored in clearly labeled, sealed containers and kept away from direct heat sources, to minimize the risk of spills, leaks, or reactions that could lead to fires or toxic exposures. Likewise, supplies like transfer papers and fabrics should be stored in a manner that protects them from contamination, damage, or degradation.

Finally, emergency procedures should be established and communicated to all who use the heat press. This includes knowing how to respond to chemical spills, fire, or exposure incidents, as well as having first aid supplies and emergency contact information readily available. Regular training and drills can help ensure that everyone knows how to act quickly and effectively in case of an emergency.

Handling chemicals and supplies safely in the context of heat press operations is a multifaceted responsibility that encompasses knowledge of the materials, proper use of PPE, adequate ventilation, safe storage practices, and preparedness for emergencies. By giving these health and safety considerations the attention they deserve, operators can protect themselves and others while also ensuring the success and sustainability of their heat press projects.

Conclusion

In wrapping up the journey of learning how to use a heat press, it's essential to revisit and consolidate the myriad aspects that contribute to mastering this versatile tool. The journey from a novice to becoming proficient with a heat press involves not just understanding the machinery but also appreciating the breadth of creativity and business opportunities it unlocks.

At its core, using a heat press effectively requires a grasp of the technical nuances—ranging from the type of press that best suits one's needs to the precise settings of temperature, pressure, and time for various materials and designs. This understanding ensures the production of high-quality, durable prints that meet the desired outcomes whether for personal projects or commercial ventures.

Beyond the technicalities, the journey emphasizes the importance of safety and proper handling of materials. It underlines how the mindful operation of a heat press, coupled with the responsible management of chemicals and supplies, not only safeguards the user's health but also extends the lifespan of the equipment and the quality of the creations. This careful approach demonstrates a respect for the craft and a commitment to maintaining a sustainable and safe working environment.

Furthermore, the exploration into the heat press world showcases the tool's versatility. From custom apparel to unique home décor, the heat press opens up endless possibilities for personalization and entrepreneurship. It encourages users to experiment with various materials and techniques, pushing the boundaries of traditional printing and design.

The economic aspect of using a heat press cannot be overstated. For entrepreneurs, it represents a cost-effective method to launch and expand a business with relatively low overhead costs. The ability to produce items on demand reduces waste and inventory expenses, making it an attractive option for small businesses and startups looking to carve a niche in the market.

Equally important is the community and resources available to those venturing into heat pressing. The shared experiences, tips, and tricks from fellow users enrich the learning journey, providing support and inspiration. This collective knowledge base is invaluable, offering insights into troubleshooting, creative ideas, and business strategies.

In conclusion, mastering the use of a heat press is a multifaceted achievement that blends technical skills, creativity, safety consciousness, and entrepreneurial spirit. It is a journey that rewards patience, experimentation, and continuous learning. For those willing to invest the time and effort, the heat press becomes not just a tool but a gateway to expressing creativity, launching businesses, and creating personalized items that bring joy and satisfaction. The key takeaways from this exploration into using a heat press are a testament to the machine's potential to transform ideas into tangible realities, making it an indispensable asset in the world of design and personalization.

Made in United States
Troutdale, OR
04/18/2024